OMNI (VOR)

ADF

ADF & OMNI (VOR) INFLIGHT EXERCISES

ARE YOU SURE OF 'WHERE YOU ARE' EVERYTIME YOU FLY?

More pilots have trouble understanding navigation than possibly any other area dealing with flying. Wouldn't it be great to know exactly 'where you are' everytime you fly using very little effort? There are systems being developed and tested now that will give you this capability. However, they are beyond the means of most of us, plus they are not yet ready for what is commonly referred to as 'general aviation.' But even these systems require you, the pilot-in-command, to think. And since thinking will always be with us, and certainly a prerequisite for professionalism, we will be looking at three very useful instruments.

Course Deviation Indicator   **Heading Indicator**   **ADF Direction Indicator**

Looking at these instruments and taking what they tell you, inserting that data into your computer (your brain), projecting that onto the screen in your head so that the picture of 'where you are' is clear; and by knowing 'where you are' as a result of that data to be able to go where you want to go whether you can see the ground or not is very important. Not just from the standpoint of today's increasing complexity of your aircraft's operating environment, but because of the safety and efficiency allowed those who understand clearly the data which the Course Deviation Indicator and the Heading Indicator or the ADF Direction Indicator and the Heading Indicator give them.

There are a lot of similarities between the Course Deviation Indicator and the ADF Direction Indicator systems of navigation. Throughout this manual we will be looking at the essentials of each and comparing their similarities.

The ADF Direction Indicator system of navigation has been with us for a long time. But because it was hard for most pilots to understand, it fell out of favor and was, for all practical purposes, replaced by the Course Deviation Indicator system of navigation. For most pilots the Course Deviation Indicator system of navigation seemed to offer an easier and more tangible form of navigation. It essentially replaced the ADF Direction Indicator system of navigation. However, even though the Omni/Course Deviation Indicator system of navigation is the most frequent type used it still presents problems to some pilots. The ADF Direction Indicator system of navigation too, is hard for most pilots to understand; more so than the Course Deviation Indicator system of navigation. Because of this and because there are a great many similarities between the Course Deviation Indicator and the ADF Direction Indicator systems of navigation I am presenting them in side-by-side illustrations with accompanying written text so that the similarities will be more clearly evident. I hope you like this format. I use it here because one of the things that has always bugged me about **all** textbooks is that the text and the illustrations to which the text referred were hardly ever in the same area where they could be easily compared.

Sesame Street, I think, has pretty well established the premise that association and repetition are vital to the learning process. It also illustrates that learning can, and should be, fun and interesting. And I think we can also say that remaining professional can, and should be, fun and interesting as well as being a great confidence builder.

In the April, 1973, issue of Approach, a Navy Safety Center Publication, CPO P. A. LaMarche in an article entitled 'Denial' said that '. . . in any survival situation, the most logical thinking person in the world can become the most illogical survivor. **This is the reason for repetition in your training.**

But bear in mind that though repetition is vital IT IS NOT VALID WITHOUT COMPLETE UNDERSTANDING.

I hope that what you read and study here brings you a better understanding of how the Course Deviation Indicator and the ADF Direction Indicator systems of navigation work and how they will, with proper use and understanding on your part, always tell you 'where you are.'

It is with the preceding thoughts in mind that I wrote this material and drew these illustrations.

My thanks to Harold Holmes for editing this manual and for his indispensable thoughts and suggestions.

INTRODUCTION

For most pilots studying the Course Deviation Indicator system of navigation and/or the ADF Direction Indicator system of navigation for the first time varies in their ease of comprehension. A lot of pilots have trouble with both systems of navigation. The two systems of navigation do vary in their complexity, but there are similarities that are worthwhile to consider so that understanding both of them comes easier.

This new type of study material is set up so that the similarities between the Course Deviation Indicator system of navigation and the ADF Direction Indicator system of navigation are more easily seen. They are presented on opposing pages with illustrations and written text for each. This makes for easy comparison and comprehension and enhances understanding of the two systems of navigation.

The text material with the accompanying illustrations and the 'in-flight' exercise sheets are not an in-depth study of the Course Deviation Indicator system of navigation or the ADF Direction Indicator system of navigation, but rather a guide to understanding how to use either of them to your best advantage.

This new type of study material also illustrates that it is not necessary to think of the Course Deviation Indicator system of navigation one way and the ADF Direction Indicator system of navigation another. Their similarities will come through very clearly as you study this material. You will note, and I will point out to you from time to time, that the Heading Indicator **and** the Magnetic Compass are indispensable to both systems of navigation. Both the Course Deviation Indicator and the ADF Direction Indicator systems of navigation rely on the Heading Indicator readings—which in turn relies on the Magnetic Compass. The Magnetic Compass is vital to the Heading Indicator so that any deviation in the Heading Indicator readings due to precession errors can be corrected by looking at the Magnetic Compass in WINGS-LEVEL FLIGHT — ONLY AND ALWAYS IN WINGS-LEVEL FLIGHT — and by re-setting the Heading Indicator according to the Magnetic Compass reading.

The similarities of the two systems of navigation will, as I said earlier, come through as you study this material. The big difference to you, the pilot-in-command, is the information that is being presented to you on the instrument panel of your aircraft by the Course Deviation Indicator and/or the ADF Direction Indicator. The Course Deviation Indicator will, when properly set, tell you with words when you are on the TO or FROM side of the Omni and by needle deflection when you are right or left of the course. And when intercepting the course, when you are at the point when you should make your turn to intercept the course. The ADF Direction Indicator in contrast only and always points to the NDB no matter where you are in relation to the NDB. It requires you to figure in your head how to go TO or FROM the NDB to which you are tuned. And also when you've reached the angle of intercept and can start your turn to intercept the course. (NDB—Non-Directional Beacon.)

In this manual you are able to study and examine numerous pictorial situations that correspond to a particular flight path. It gives you all the situations you'll ever encounter when flying. The illustrations on the opposing pages are the same except for the readings on the dials of the Course Deviation Indicator and the ADF Direction Indicator. But even though the readings are different both tell you that you are in the same location in regards to the Omni or NDB station on the ground to which you are tuned. There is a fold-out page to show you an over-all view of the flight path of the aircraft used herein for illustration purposes and that you can follow as you study.

Understanding how the Course Deviation Indicator and the Heading Indicator relate to each other. How the ADF Direction Indicator and the Heading Indicator relate to each other when using their respective systems of navigation can give you the information you need to know 'where you are' and from there 'where you want to go.'

There are two sheets. One for the Course Deviation Indicator system of navigation and one for the ADF Direction Indicator system of navigation. They accompany this manual and can be used to fly the same course in the air that you studied on the ground. You can also use them when you practice in a fixed or moveable simulator. (Variations of the flight path sheets are available from the Aviation Department of Haldon Books for a nominal fee.)

Always use a safety pilot when you fly these practice sessions in your aircraft, whether you use an instrument hood or not. Study this material thoroughly then practice them until you're able to 'see in your mind' where you are as you fly. Because until you can actually 'see in your mind where you are' by looking at the Course Deviation Indicator and the Heading Indicator or the ADF Direction Indicator and the Heading Indicator you'll have a hard time understanding either the Course Deviation Indicator or the ADF Direction Indicator system of navigation.

REMEMBER THAT IT IS VITAL THAT WHAT YOU'VE STUDIED BE VALIDATED IN YOUR MIND.

Using both the Course Deviation Indicator system of navigation and the ADF Direction Indicator system of navigation requires thought on your part because — IN THE FINAL ANALYSIS WHEN YOU'RE PILOT-IN-COMMAND, THE DECISION AS TO 'WHERE YOU ARE' RESTS WITH YOU.

Study the material and fly the exercises as suggested and you will be building confidence in yourself as pilot-in-command and when you say to yourself, 'I know where I am,' YOU REALLY WILL KNOW.

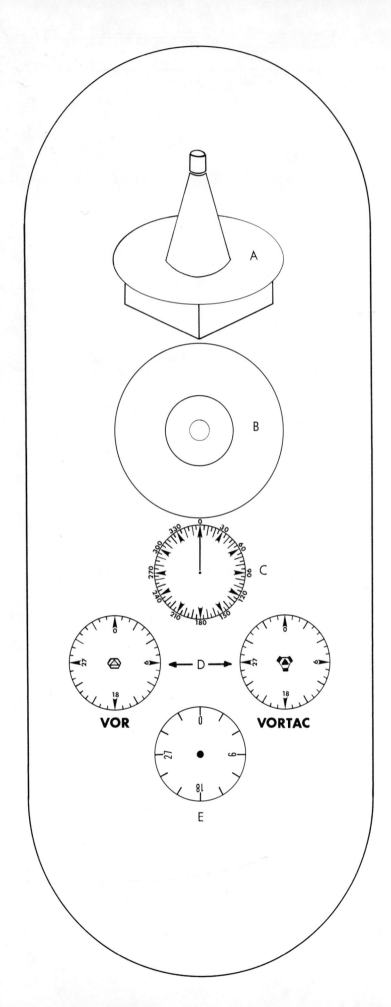

OMNI NAVIGATION

There are three very important aspects of Omni navigation that you must learn in order to become and remain professional. They are **ORIENTATION** (where am I?), **INTERCEPTION** (how to intercept the proper course to take you TO or FROM the Omni), and **TRACKING** (the art of staying 'on course' no matter what the direction of the wind). These three must be studied and practiced so that they will become 'second nature' because no matter what happens, whether it be low visibility, complete cloud cover, darkness, or stress, you want to be able to know 'where you are' and from there go to 'where you want to go,' be it your destination or an alternate airport. And you want to be able to do it **KNOWING THAT YOU KNOW EXACTLY 'WHERE YOU ARE' AND 'WHERE YOU'RE GOING!'**

Before we discuss ORIENTATION, INTERCEPTION or TRACKING let's spend a few minutes discussing the Omni station (hereinafter referred to as simply the 'Omni') and the two very important instruments you'll be studying and using throughout the text and the 'in-flight' exercises; namely the Course Indicator and the Heading Indicator.

WHAT DOES THE OMNI LOOK LIKE?

A Well, as you fly and look down on the Omni from your aircraft it looks similar to the illustration on your left. Sort of a modern type of wigwam on top of a circular disc on top of a square house that houses the electronic innards of the Omni.

B If you were to look directly down on top of the Omni it would look something like this. Three circles, one on top of the other.

C If you look at the visual flight rules (VFR) charts (called Sectionals) the Omni will appear to be something like this.

D If you look at the instrument flight rules (IFR) charts (called Enroute Low Altitude or Enroute High Altitude charts) the Omni will appear like one of these two depending on whether it is a VOR or a VORTAC Omni. The first is a VOR Omni that sends out Morse code signals only. The second is a VORTAC Omni that sends out Morse code and a recorded voice. More on that later.

E This is how the Omni looks in the text material and the pictorial part of the 'in-flight' exercises.

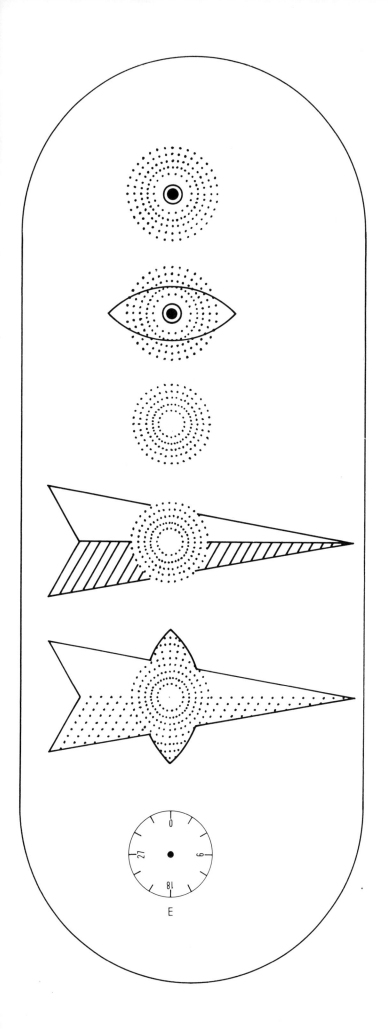

NDB NAVIGATION

The ADF Direction Indicator system of navigation is comprised of the same three considerations as the Course Deviation Indicator system of navigatiton; namely ORIENTATION ('where am I in relation to the station to which I'm tuned?'), INTERCEPTION (how do I intercept the course TO or FROM the station to which I am tuned?), and TRACKING (the art of staying 'on course' no matter what the direction of the wind). These three must be studied and practiced so that they become 'second nature,' because no matter what happens, whether it be low visibility, complete cloud cover, darkness or stress you want to be able to know 'where you are.' And from that point to go to 'where you want to go,' be it your destination or an alternate airport. And you want to be able to do it knowing THAT YOU KNOW EXACTLY 'WHERE YOU ARE,' 'WHERE YOU'RE GOING,' and 'HOW YOU'RE GOING TO GET THERE.'

Before we discuss ORIENTATION, INTERCEPTION OR TRACKING let's discuss the NDB (Non-Directional Beacon). What it looks like and where it is located and the two very important instruments you'll be looking at as you study this material on the ADF Direction Indicator system of navigation.

WHAT DOES THE NDB (NON-DIRECTIONAL BEACON) LOOK LIKE?

Unless you're standing right next to the NDB you'd probably never see it. In contrast to the Omni ground station the NDB cannot be seen from the air. When you use the Flight Path Exercise Sheets you will have to locate it on the map you're using, then use some landmark nearby for a position marker. When you look at a VFR (Visual Flight Rules) chart — also called a Sectional — you see them scattered across the land, although with less frequency than the Omni ground stations. It will appear on an Enroute Low Altitude Chart the same way.

The illustrations to the left are how they appear on the appropriate charts. For this manual the NDB will be depicted as it on the Omni illustrations, i.e., the last drawing on the list.

The NDB may or may not be located on the airport proper. If you're landing or shooting landings at an airport with an ILS (Instrument Landing System) approach you will find an NDB located on the approach path to the airport designated as shown in the illustration opposite. It may say LOC for Localizer and it may not. But on an Approach Chart (sometimes called Approach Plate) it will be designated LOM (Localizer Outer Marker) and it will be shown as such and located on the end of the runway.

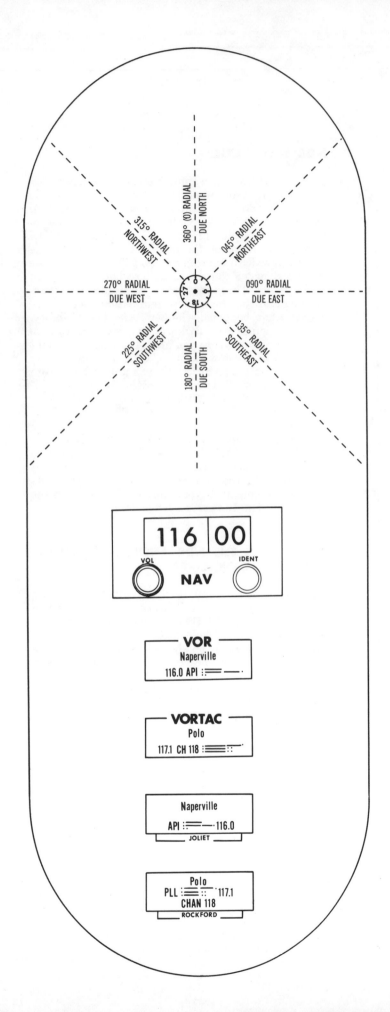

WHAT DOES THE OMNI DO?

The word **omni** is derived from the Latin word OMNIS meaning **all**. Omnidirectional then means all directional. The purpose of the Omni is to send out radio signals in all directions of the compass. The compass is a complete circle containing 360 degrees. There is one signal for each degree of the compass, therefore the Omni sends out 360 radio signals, one in each direction. Each radio signal that emanates from the Omni is named for the number of degrees it is from 'magnetic' North, counting clockwise, and for the direction in which it is pointing. For example, the 090° radial is 90 degrees to the right of 'magnetic' North and is pointing to what is commonly referred to as 'due-East.'

In the accompanying illustration there are eight (8) radials showing their position and the direction in which they are pointing.

HOW DO I PICK UP THE SIGNALS FROM THE OMNI?

By turning the frequency selector knob on the navigation side of your radio receiver until the proper frequencies of the Omni you have 'selected' appear in the opening. Once you have the proper frequencies you must turn the volume knob so that you can listen for the Morse code, or the Morse code and a recorded voice to make sure you are tuned to the Omni you want. **YOU MUST LISTEN. THE CALL NUMBERS ALONE WILL NOT ENSURE YOU'RE TUNED IN TO THE OMNI YOU WANT.**

If it is a VOR Omni it will send out Morse code signals only. Those signals, made up of dots and dashes, will appear in the box on the Sectional chart containing the name of the VOR.

If it is a VORTAC Omni it will send out Morse signals plus you will hear a recorded voice say (using the illustration to your left) 'Polo Vortac.' The voice and the Morse code are repeated every few seconds so there's no waiting and you are sure of being tuned to the Omni you want. The illustration shows how it appears on the Sectional charts.

This is how the same VOR and VORTAC boxes would appear on the Enroute Low Altitude Charts.

The name below each box is the Flight Service Station (FSS) that controls each VOR or VORTAC.

The Chan 118 on the Polo VORTAC is Channel 118 and is used by the military for instrument approaches.

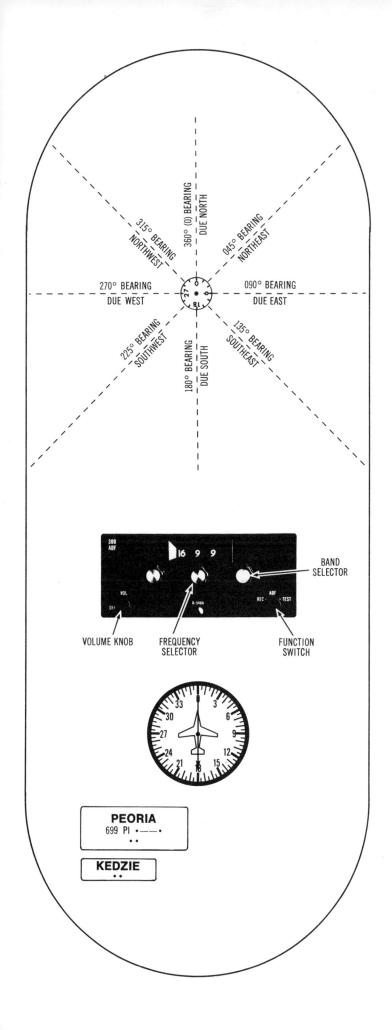

WHAT DOES THE RBn (RADIO BEACON) NDB (NON-DIRECTIONAL BEACON) TRANSMITTER LOOK LIKE?

The NDB transmitter is described as being non-directional in nature, and when compared to the Omni transmitter, this is true. However, the fact remains that the NDB transmitter does have radials just like the Omni and even though they are called bearings they still play the same roles as the radials from an Omni transmitter. If you will think of the NDB transmitter as having radials that are indexed to magnetic north (just like the Omni transmitter) it will greatly facilitate your orientation as to 'where you are' in relation to the NDB transmitter to which you are tuned.

HOW DO I PICK UP SIGNALS FROM THE NDB?

The illustration to your left depicts the more modern NDB receiver in that it has an LED (Light Emitting Diode) face rather than a printed one. The operation is the same however, so the first thing you do is to turn the function switch to REC. Then turn the band selector switch to the frequency range of the NDB transmitter to which you want to tune. Next turn the volume (VOL) control knob well up so you can hear the aural sound that will be coming from the station after you turn the selector (SEL) knob to the identifying numbers that will appear in the rectangular face. And as with the Omni make sure you have the correct station by listening for the appropriate Morse Code signal from that particular NDB to which you are tuned. Once you are sure you have correctly identified the station turn the function switch to ADF. It is at this time that you will see the ADF Direction Indicator needle move and point to the NDB transmitter to which you are tuned.

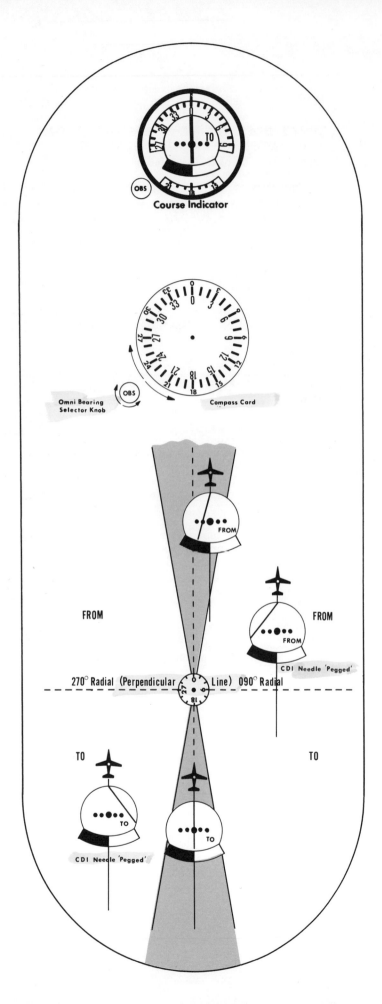

NOTE: Throughout the rest of this manual and the Flight Path Exercise Sheets the NDB transmitter will be referred to as simply the NDB.

WHAT INSTRUMENTS DO I USE WITH THE NAVIGATION RECEIVER?

The Course Indicator is the only instrument directly connected to the navigation side of your radio receiver. The Course Indicator displays the necessary data for you to make a decision as to 'where you are' in relation to the Omni to which you are tuned.

HOW DO I INTERPRET WHAT IT IS TELLING ME?

In order to help you better understand what the Course Indicator is telling you, I've separated it into two parts.

All the parts are inter-connected, either electronically or mechanically. By turning the Omni Bearing Selector (OBS) knob, you rotate the Omni Bearing Compass Card, which in turn affects the TO-FROM indicator reading and the Course Deviation Indicator (CDI) needle reading.

For purposes of this illustration the Course Indicator is set to 0 (360°). Which means that your 'selected course' (TO and FROM the Omni) is 0 (360°). The 'selected course' will always appear at the center of the top opening on the Course Indicator dial in the text and the 'in-flight exercises.' (Some Course Indicators are set up in the reverse. **Make sure you know how yours operates.**) The number that appears at the center of the bottom opening on the Course Indicator dial is always the reciprocal (opposite) of the 'selected course' number. (By rotating the book you will note this to be true in all cases on the Omni Bearing Compass Card.) The compass card is printed as illustrated so that the 'selected course' number and its reciprocal are easily readable.

The shaded area emanating from the center of the Omni is always 10° either side of your 'selected course' and its reciprocal. If your aircraft is **inside** the shaded area (i.e., within 10° to the right or 10° to the left of your 'selected course' — no matter what the direction of travel) the CDI needle will indicate this by being in a position somewhere between the large dot in the center and the 'pegged' position. And it will be either right or left depending on whether your aircraft is right or left of your 'selected course.'

If your aircraft is **outside** the shaded area the CDI needle will be 'pegged' (i.e., it is as far as it will go either right or left; depending on whether you're right or left of your 'selected course.')

And no matter where you are in relation to the Omni there will always be a line that is perpendicular to your 'selected course' that runs through the center of the Omni and this is the point at which the TO-FROM readings change. In the illustration that perpendicular line is represented by the 270° and 090° radials.

HOW DO I INTERPRET WHAT THE ADF DIRECTION INDICATOR IS TELLING ME?

ADF means Automatic Direction Finding. Which means that once you are tuned to the correct NDB the needle of the ADF Direction Indicator will automatically find the direction from which the electronic signal is generated. It will point in that direction no matter what your magnetic direction of flight. The ADF Direction Indicator needle does not point in a magnetic direction. THE ONLY DIRECTION THE ADF DIRECTION INDICATOR NEEDLE POINTS IS THE DIRECTION OF THE NDB AS IT RELATES TO THE NOSE OR THE TAIL OF YOUR AIRCRAFT. THAT'S VERY IMPORTANT. When you look at your ADF Direction Indicator and you note that the needle is pointing to a position that is just ahead of your left wing — point in that direction and say 'it's out there' — that will help you visualize where it is. Or if it is pointing to a position that is behind your right wing — point in that direction and say 'it's back there.' Visualizing where the NDB is in relation to the nose or tail of your aircraft is very important. As you study this material some of this will be easy to understand BECAUSE IT IS LAID OUT FLAT AND IT LOOKS SO RIGHT AND SIMPLE. But when you are in your aircraft or a flight simulator the ADF Direction Indicator, and all other instruments, are in an up-right position. It's confusing because no matter where the ADF Direction Indicator needle points except when pointing to 27 (your left wing tip) or to 9 (your right wing tip), it either points in an UP or DOWN direction. And that ain't where the NDB is! It is either (1) directly in front of you, (2) somewhere between the nose of your aircraft and your right wing tip, (3) somewhere between your right wing tip and the tail of your aircraft, (4) directly behind you, (5) somewhere between your left wing tip and the tail of your aircraft, or (6) somewhere between the nose of your aircraft and your left wing tip.

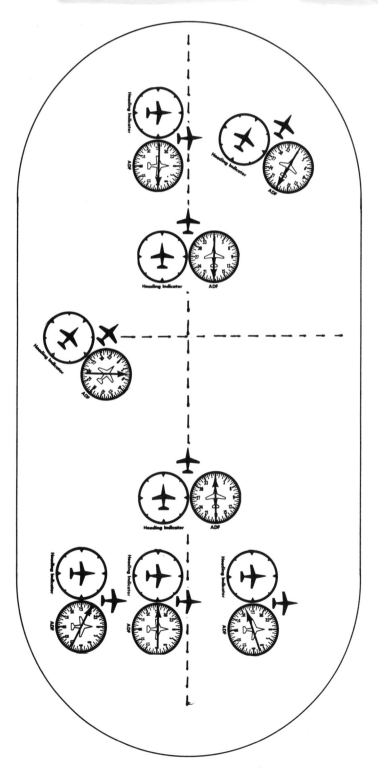

The numbers on your ADF Direction Indicator dial are, **at this time,** completely useless and for the moment do not give you any worthwhile magnetic direction information. In the accompanying illustration I've purposely left the numbers off the Heading Indicator dial to illustrate a point. In all of the illustrations the ADF Direction Indicator needle is pointing to the station and again, THE ONLY THING IT IS TELLING YOU IS WHERE THE STATION IS IN RELATION TO THE NOSE OR TAIL OF YOUR AIRCRAFT — AND REMEMBER **THAT IS VERY IMPORTANT.** Knowing where the NDB is in relation to the nose or tail of your aircraft helps you visualize your location in relation to the NDB. In these illustrations the dial is laid out flat which makes it easier to understand. During WW II some of the British aircraft had their compasses lying flat rather than on the vertical plane as they are on the instrument panel. It was placed that way for easy comprehension. Directional orientation is certainly easier but it does take up much needed space and it also requires more head movement which is not good in IFR (Instrument Flight Rules) conditions.

So remember it is very important that you first physically point in the direction of the NDB in relation to the nose or tail of your aircraft so you can visualize where it is. On a later page we will discuss how the numbers on the ADF Direction Indicator dial, coupled with the Heading Indicator reading, tell you what magnetic direction to fly to take you TO or FROM the NDB to which you are tuned.

NOTE: Due to the size of the illustrations the ADF Direction Indicator needle is not actually pointing **to** the NDB but if it were in the aircraft it would be. The needle is placed where it is in relation to the position of the aircraft in the illustration.

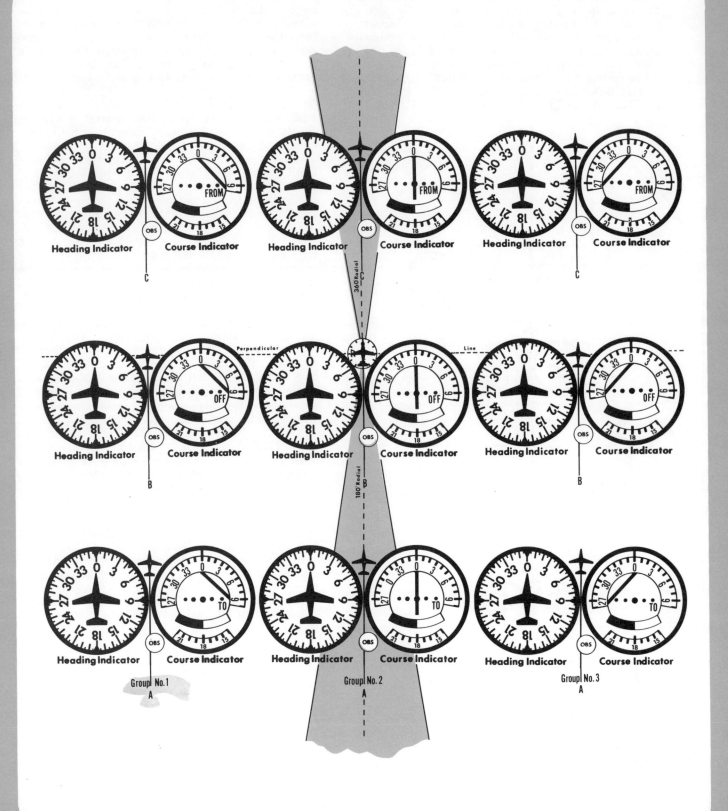

WHAT THE OMNI COURSE DEVIATION INDICATOR READINGS WOULD BE WITH THE CORRECT DATA INSERTED

In the preceding pages we've looked at the Omni and the Course Deviation Indicator in some detail. Let's now discuss both by putting them on one page showing how different readings occur and what they mean. I've added the Heading Indicator simply to show the aircraft heading in each illustration. Nine aircraft are illustrated, but we will assume that there is only one aircraft and it will be placed in the nine different positions indicating how the Course Indicator readings would appear in each case in a 'no-wind' condition, i.e., the wind is calm.

The Course Indicator in all instances is set to 0, i.e., the 'selected course' is 360°, due-North. Before the aircraft crosses the perpendicular line, the Course Indicator will read **TO**. When the aircraft is **on** the perpendicular line (this includes being **on top** the Omni too!) the Course Indicator will read **OFF** and after the aircraft crosses the perpendicular line the reading will change to **FROM**.

Now let's discuss each individual group of illustrations.

NOTE: The **'magnetic heading'** on the Heading Indicator and the 'selected course' on the Course Indicator are the **same in all positions** of your aircraft in these illustrations whether you are right or left of the 'selected course' or whether you are on the TO side of the perpendicular line, crossing it, or on the FROM side of the perpendicular.

GROUP #1

(A) The CDI needle is pointing to the right, indicating that the 'selected course' is to your right. The CDI needle is 'pegged' indicating that you are **OUTSIDE THE CDI NEEDLE SENSITIVITY AREA.** You have a TO reading because you are on the TO side of the perpendicular line.

(B) The CDI needle still indicates your 'selected course' to be to your right. However, the TO-FROM reading indicates OFF. OFF because you are going from the TO side to the FROM side of the perpendicular line. Everything else remains the same.

(C) You're still to the left of your 'selected course' therefore the CDI needle still points to the right.

You have a FROM reading because you are on the FROM side of the perpendicular line represented here by the 270 and 090° radials.

GROUP #2

(A) Now the CDI needle is centered because you are 'right on course.'

You're on the TO side of the perpendicular line. All other settings are the same as GROUP #1.

(B) You're 'on top' the Omni. The TO-FROM reads OFF again for the same reason as before, i.e., you are crossing the perpendicular line.

The perpendicular line extends outward in both directions from the Omni and depending on the altitude of your aircraft the length of time the Course Indicator says OFF will vary each time you cross it. **WHEN YOU ARE TRACKING TO AND FROM THE OMNI YOU SHOULD ALWAYS TRY TO FLY EXACTLY OVER THE 'TOP' OF THE OMNI. YOU'LL KNOW WHEN YOU DO THIS BECAUSE YOUR CDI NEEDLE WILL REMAIN STATIONARY.**

(C) Still 'on course,' but now you've crossed the Omni so you are on the FROM side of the perpendicular line.

GROUP #3

(A) This position is like (A) of GROUP #1 except you're on the right side of the 'selected course,' therefore the CDI needle points to your left, which means that to get back on your 'selected course' you must turn left.

You are still on the TO side.

(B) The CDI needle says the 'selected course' is to your left. And again you're crossing the perpendicular line and that's why the Course Indicator reads OFF.

(C) You've crossed the perpendicular line so now you're on the FROM side again. The CDI needle is 'pegged' left because you are out of the 10° 'selected course' area.

NOTE: In order to get 'on course' in the illustrations depicted in GROUP #1 and 3 you would, in all cases, **TURN TOWARD THE NEEDLE,** i.e., turn your aircraft in the direction in which the needle is pointing. The old adage 'turn toward the needle' has been used by pilots for years and it is still valid **AS LONG AS YOU KNOW 'WHERE YOU ARE' AND HAVE INSERTED THE CORRECT DATA INTO THE COURSE INDICATOR.**

What happens when you dial in the wrong data into the Course Indicator? Turn the page . . .

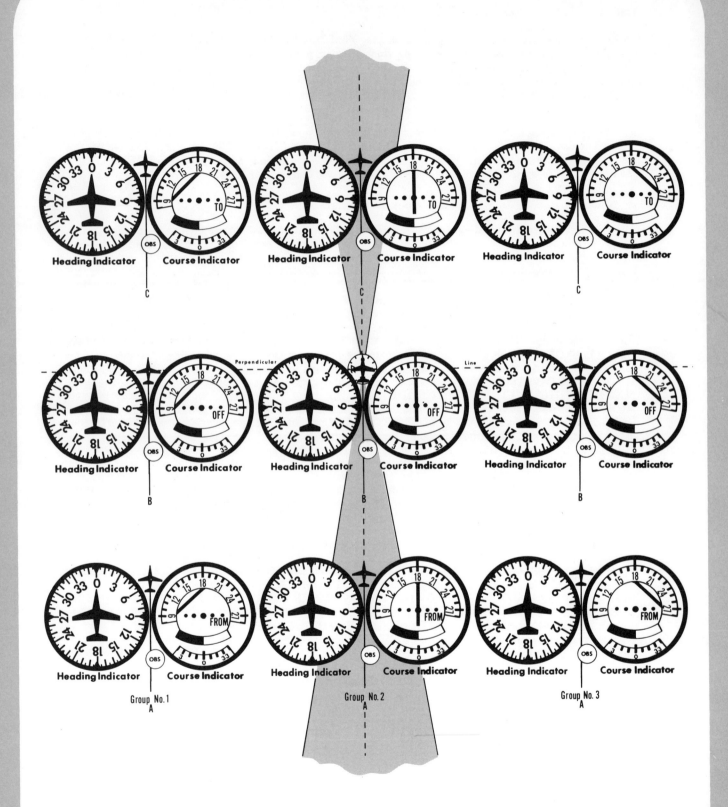

WHAT THE OMNI COURSE DEVIATION INDICATOR READINGS WOULD BE WITH THE INCORRECT DATA INSERTED

In the preceding illustration and discussion concerning the Course Indicator with the '**correct data**' inserted the 'selected course' was 0 (360°, due-North), therefore that number, (0), was at the center of the top opening on the Course Indicator dial. When your aircraft went from (A) to (B) to (C) the TO-FROM reading went from TO to OFF to FROM. And when you were to the right of your 'selected course' the CDI needle said the 'selected course' was to your left **(THINK ABOUT THAT STATEMENT A LITTLE TO MAKE SURE YOU AGREE WITH IT)** and that is correct. And you also remember that the 'magnetic heading' was the same at the 'selected course.'

Now, **BY JUST CHANGING THE 'SELECTED COURSE' ON THE COURSE INDICATOR YOU WILL GET A COMPLETE REVERSAL OF ALL THE READINGS.** This is what I term inserting the wrong data into the Course Indicator. The wrong data being a 'selected course' of 18 (180°, due-South), when what you want, and should have, based on your relationship to the Omni and the desire to go TO the Omni, as previously shown is a 'selected course' of 0 (360°, due-North).

There are two things in this particular case that will remain the same even though you've changed the 'selected course' to 18 (180°, due-South). One is the perpendicular line; it is still represented by the 270 and 090° radials. The second is that the Course Indicator will still indicate OFF when you cross the perpendicular line.

GROUP #1

(A) The first reversal that is evident is the CDI needle. The old adage 'turn toward the needle' will not work here, especially if you wanted to get back 'on course.' The complete reversal we discussed a b o v e goes for the CDI needle even to the point that it is 'pegged,' even though to the wrong side.

The TO-FROM says you're on the FROM side of the perpendicular line. It's easy to see from t h e illustration that you're not, but would **YOU SEE IT THAT EASY WHILE YOU WERE FLYING?**

(B) You're crossing the perpendicular line and you have an OFF reading just like in the previous illustrations. **YOU ALWAYS GET AN 'OFF' READING WHEN YOU CROSS THE LINE THAT IS PERPENDICULAR TO YOUR 'SELECTED COURSE' WHETHER YOU HAVE THE 'CORRECT' OR 'INCORRECT' DATA INSERTED IN YOUR COURSE INDICATOR.**

(C) You're not on the TO side of the perpendicular line are You? You're going away from the OMNI. A complete reversal of what it should be.

GROUP #2

(A) As long as you're 'on course,' no matter whether you have the correct or incorrect data in the Course Indicator, the CDI needle will center. The only confusing thing again is the fact that the TO-FROM says you are on the FROM side. Again, you know by looking at the illustration that this is false, **BUT UNDER STRESS WHILE FLYING WOULD YOU BE ABLE TO TELL?**

(B) You're 'on top' the Omni which is the same as crossing the perpendicular line, therefore you have an OFF reading on the Course Indicator.

(C) All of a sudden you cross the Omni and you're on the TO side, going AWAY FROM THE OMNI. Again, a complete reversal.

GROUP #3

(A) The CDI needle says the 'selected course' is to your right. Would you turn right to get 'on course?' **WOULD YOU, UNDER STRESS, NOTICE IF YOU HAD INSERTED THE WRONG DATA INTO YOUR COURSE INDICATOR? AND FLY AWAY FROM THE COURSE INSTEAD OF TOWARDS IT?**

Would you fly TO the Omni with a FROM reading? It's easy to do with the wrong data in the Course Indicator.

(B) **CROSSING THE PERPENDICULAR LINE ALWAYS GIVES AN 'OFF' READING.**

(C) You would never fly away from the Omni to a TO reading WOULD YOU? With the wrong data inserted it's mighty easy!

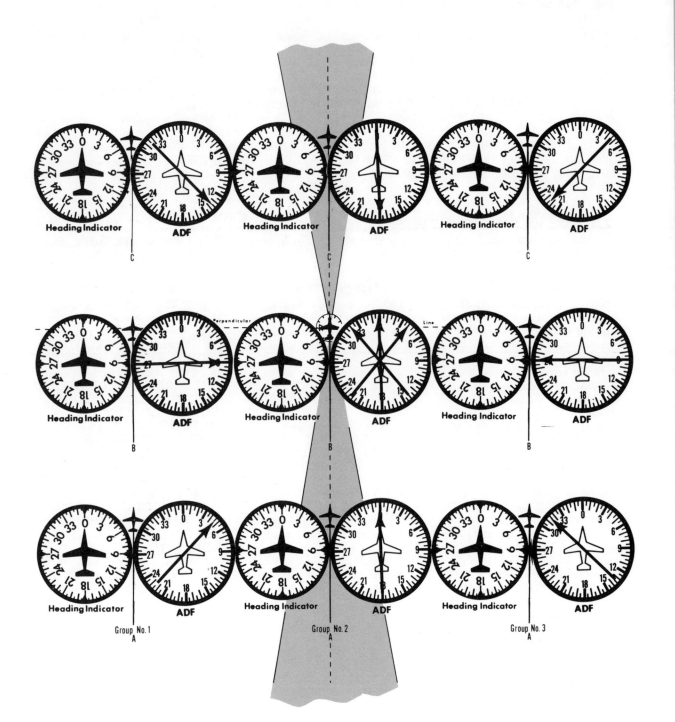

WHAT THE ADF DIRECTION INDICATOR READINGS ARE WHEN TUNED TO THE NDB

As you probably noticed this is a rather straight forward illustration in that all readings are such that the needle indications are the same as the magnetic directions you would turn to go TO the NDB. YOU WOULD HAVE TO BE IN THIS EXACT SITUATION BEFORE THIS WOULD BE TRUE. Read on — we'll get to the mathematical calculations soon and then you'll understand the ADF Direction Indicator system of navigation better.

On the four preceding pages we examined the Omni Course Deviation Indicator's reading of nine aircraft as they would appear in each of their locations as depicted in the illustrations. On two of the pages the correct data was inserted and on the other two the incorrect data was inserted. In these nine illustrations we will examine the readings of nine aicraft in the same location, as regards a station, and see how they differ. Because of the operation of the ADF Direction Indicator you cannot insert wrong data. The needle of the ADF Direction Indicator ALWAYS POINTS TO THE NDB TO WHICH YOU ARE TUNED NO MATTER WHERE YOU ARE IN RELATION TO THE NDB.

I've kept the Heading Indicator in this illustration simple to show the aircraft heading in each location.

As with the Course Deviation Indicator illustrations these ADF Direction Indicator illustrations are presented in a no-wind condition, i.e., the wind is calm.

The ADF receiver in all cases is tuned to the station, in the center of the illustration. I've left the perpendicular line in the illustration to show that this is the point where the needle of the ADF Direction Indicator points either to your right wing tip (9) or your left wing tip (27). On the Course Deviation Indicator it's where it goes from TO to FROM. If you will look at the ADF Direction Indicator as having this imaginary perpendicular line **too** and that when your ADF Direction Indicator needle is pointing anywhere from ahead of your right wing tip (9) to ahead of your left wing tip (27) you'll know that you're on the TO side of the station to which you're tuned. And if the needle points to either 9 or 27 you're right on the perpendicular line. And if it points anywhere in the area from the back of your right wing tip (9) to the area in back of your left wing tip (27) you'll know that you're on the FROM side of the NDB to which you are tuned.

GROUP #1

(A) The ADF Direction Indicator needle is pointing to 3. It is pointing in the area above your right wing tip (9), therefore you are on the TO side and a turn to 30° would take you TO the NDB. In this particular instance it also tells you that you are the **southwest quadrant.** The tail of the ADF Direction Indicator needle is in the southwest quadrant.

(B) Here the station is directly off your right wing tip (9) and you are on the perpendicular line going from the TO side to the FROM side. As you fly this particular path the needle will move at a steady pace from the area above your right wing tip to the area below your right wing tip.

(C) Here you've entered the **northwest quadrant** (note the tail of the ADF Direction Indicator needle). The NDB is below (or behind) your right wing tip and you are going FROM the NDB.

GROUP #2

(A) You're right 'on course' TO the station. It is right in front of you, over the nose of your aircraft. (IT'S NOT UP IN THE AIR WHERE THE NEEDLE IS POINTING.)

(B) When you are 'on top' the NDB the needle will become very irratic and depending on your location, i.e., either just to the right or just to the left on the NDB, the needle will either swing to the right or to the left and point to the tail (18) of your aircraft. THIS IS ONE TIME WHEN YOU MUST KEEP YOUR 'COOL' BECAUSE THE IRRATIC NEEDLE MOVEMENT CAN CREATE SOME ANXIOUS MOMENTS IF YOU LET IT.

(C) At C you're still 'on course' but you are going FROM the NDB. The needle of the ADF Direction Indicator is pointing right at the tail of your aircraft (18). (IT IS NOT DIRECTLY BELOW YOU.) You're on the FROM side of the perpendicular line now.

GROUP #3

(A) The NDB is in the area between the nose of your aircraft (0) and your left wing tip (27). You are on the TO side of the NDB in the **southeast quadrant.** Here the tail of the needle is pointing to that particular quadrant. To go TO the NDB you would turn left to a heading of 330°. The ADF Direction Indicator is pointing to 33.

(B) Here you are to the right of the NDB. The needle is pointing to 27 so the station is off your left wing tip. You are also crossing the perpendicular line. And as I mentioned in (B) of GROUP #1 the needle moves at a steady pace from above your left wing tip to below your left wing tip anytime you are to the right of the NDB to which you are tuned.

(C) Now you're on the FROM side of the station going away FROM the NDB. The needle is pointing to the area between the left wing tip (27) and the tail (18) of your aircraft.

The time interval from the first indication of station proximity to positive station passage varies with altitude — from a few seconds to three (3) minutes. Initial station passage is positively determined when the ADF Direction Indicator needle moves through the wing tip position (27 or 9 on the ADF Direction Indicator dial). This usually occurs shortly after your aircraft has actually passed the NDB. Timing should begin at this instant regardless of further oscillations.

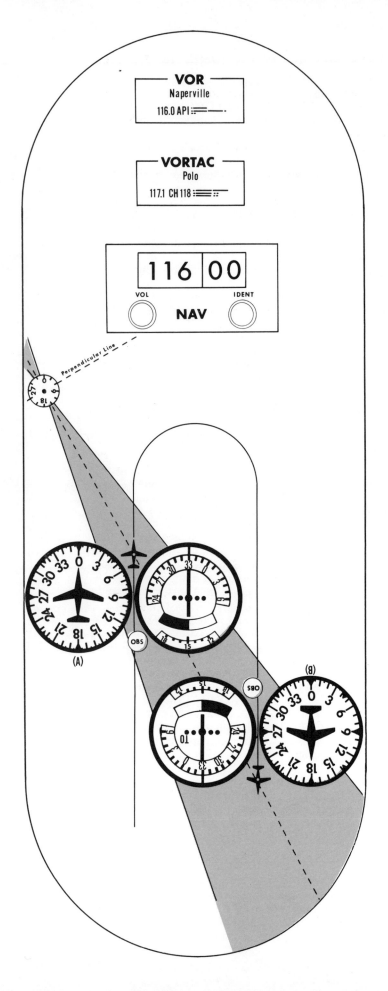

ORIENTATION & QUADRANTS

Orientation — 'where am I?', one of the three aspects of Omni navigation, is the one we'll consider first since you must know 'where you are' before you can decide how to get to 'where you want to go.'

We'll be discussing the use of the Course Indicator and the Heading Indicator and how by using them together you can easily tell 'where you are.'

There are a few basic steps that are most helpful when orientating yourself. The first would be looking on your map for an Omni (VOR or VORTAC) you think is near 'where you **think you are.**'

1. Get the call numbers from the map for the Omni you feel is near your present position. It will be either a VOR or VORTAC Omni.

2. Dial those numbers into the navigation side of your radio receiver by turning the 'ident' knob until those numbers appear in the opening. To make sure you have the Omni you want, turn the volume control up and listen for the Morse code (VOR) or the Morse code **and** voice (VORTAC). **DO NOT RELY ON THE NUMBERS ALONE.**

3. Center the CDI needle with **a** TO reading. Look first for the TO reading and once you have that center the CDI needle. You want to know the course 'inbound' TO the Omni even though you may not want to go to that particular Omni. **If you check to see what radial you're on by looking for a FROM reading you must remember that you've put the Course Indicator 'reversal symptoms' to work and unless you are aware of what you're doing you could easily get mixed-up and turn away from the 'selected course' instead of towards it. LOOK FOR THE 'TO' READING.** If you decide to go TO the Omni to which you are tuned, everything is already set-up. If you want to know what **'radial'** you're on it is **always the reciprocal of your Course Indicator setting,** providing of course that you have a TO reading.

4. Once you know the course TO the Omni you have to know your present 'magnetic heading,' i.e., in what direction am I flying right now? This is where visualization is important. You have to be able to visualize your position in relation to the Omni to which you are tuned based on the information given you by the Course Indicator and the Heading Indicator. The Course Indicator tells you the course TO the Omni and the 'radial' you are on. The Heading Indicator tells your direction of flight.

In both cases in the illustration to your left the OBS knobs have been rotated until there was a TO reading and the CDI needle centered. If you wanted to go TO the Omni your 'course' in both cases would be 330° (Northwest). However, at (A) you would turn left and at (B) you would turn right. That's why it's so important to 'visualize where you are.' In either case you are on the 150° radial in the Southeast quadrant.

At position (A) your 'magnetic heading' is 0 (360°) (due-North) and Position (B) your 'magnetic heading' is 18 180° (due-South). At both (A) and (B) you are on the TO side of the perpendicular line represented by the 240 and 060° radials.

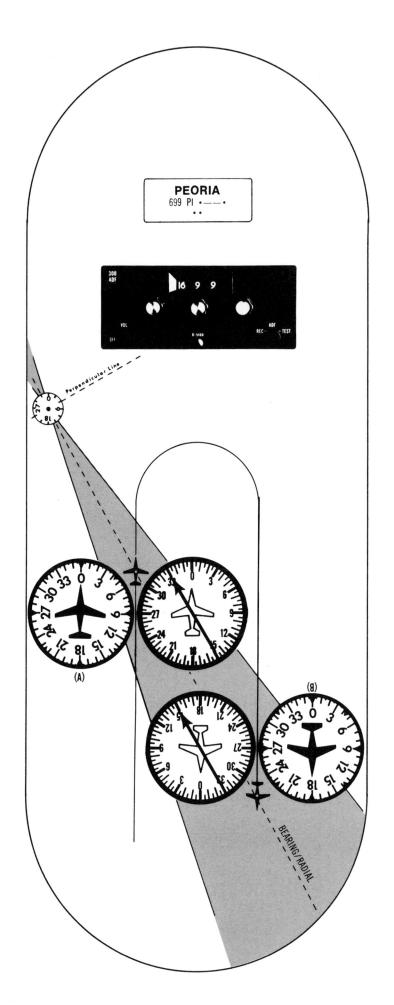

ORIENTATION & QUADRANTS

Orientation — 'where am I?,' one of the three aspects of ADF Direction Indicator navigation, is the one we'll consider first since you must know 'where you are' before you can get to 'where you want to go.'

We'll be discussing the use of the ADF Direction Indicator and the Heading Indicator and how by using them together you can easily tell 'where you are.'

There are a few basic steps that are most helpful when orientating yourself. The first would be looking on your map for an NDB you think is near 'where you think you are.'

1. Get the call numbers from the map for the NDB you feel is nearest your present position.

2. Dial those numbers into your ADF receiver by turning the Frequency Selector Knob until those numbers appear in the opening. And as with the Omni make sure you are tuned to the correct NDB, turn the volume control up and listen for the appropriate Morse Code signal. DO NOT RELY ON THE NUMBERS ALONE. Once you are sure you have correctly identified the NDB, turn the function switch to ADF. It is at this time that you will see the ADF Direction Indicator needle move and point to the NDB to which you are tuned.

3. By looking at the needle make your decision as to the location of the NDB as it relates to the nose or tail of your aircraft. Once you know where the NDB is, in relation to the nose or tail of your aircraft, you have to check to see what your present magnetic heading is, i.e., in what direction am I flying? (In this illustration it is north.) Here's where visualization is important. At (A) the NDB is to the left of the nose of your aircraft. By superimposing the needle of the ADF Direction Indicator to the dial (face) of the Heading Indicator you can see that a heading of 330° would take you to TO the NDB. In this case the needle points to 33 (330°). But look at (B); the needle points to 15 (150°). You certainly wouldn't turn to a heading of 150° to go TO the NDB, but by superimposing the ADF Direction Indicator needle onto the Heading Indicator again you see that the course TO the NDB is 330°. The needle tells you that the NDB (B) is behind your right wing.

19

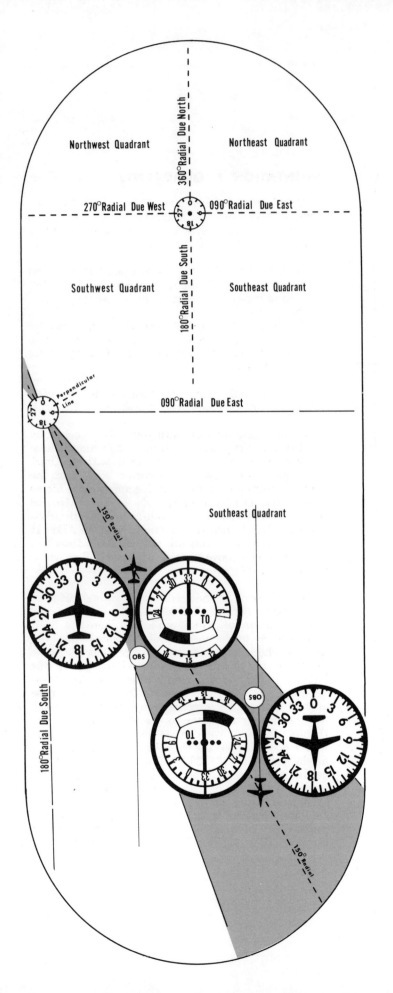

5. Another valuable tool of orientation is to know what quadrant (Northeast, Southeast, Southwest or Northwest) you are in while checking your **course** TO the Omni and the **radial** you are on. Knowing what quadrant you're in helps you 'visualize' your position in relation to the Omni. And it helps you in deciding what direction and heading to turn to.

The NORTHEAST QUADRANT encompasses all 'radials' from the 0° (360°, due-North) radial to the 090° radial (due-East). The SOUTHEAST QUADRANT covers all the radials from 090° to 180° (due-South). The SOUTHWEST QUADRANT runs from the 180° radial (due-South) to the 270° radial (due-West). The NORTHWEST QUADRANT then completes the circle by covering the area from the 270° radial (due-West) to 0° radial (360° due-North).

HOW DOES KNOWING WHAT QUADRANT I'M IN HELP ME TO ORIENT MYSELF IN RELATION TO THE OMNI TO WHICH I AM TUNED?

Looking at the accompanying illustration we see that your course TO the Omni is 330° (Northwest) and you're on the 150° radial (Southeast). By knowing you are on the 150° radial you know you're in the Southeast Quadrant, the area between the 090° radial (due-East) and the 180° radial (due-South); **therefore you are Southeast of the Omni.** To help you visualize your position in relation to the Omni you have the Heading Indicator. Visualize yourself in the SOUTHEAST QUADRANT then look at your Heading Indicator to determine the 'magnetic direction' in which you are flying. In this case it is 0° (360° due-North). The 'course' TO the Omni is 330° (Northwest) therefore to go TO the Omni you would turn left to a heading of 330. **MENTALLY OR PHYSICALLY POINTING IN THE DIRECTION OF THE OMNI WILL HELP YOU VISUALIZE 'ITS' POSITION.**

In the second illustration you are on the 150° radial again, but your heading has now changed to 180° (due-South); therefore to go TO the Omni you would turn right to a heading of 330°. (We'll talk about the angle of intercept later in the text.)

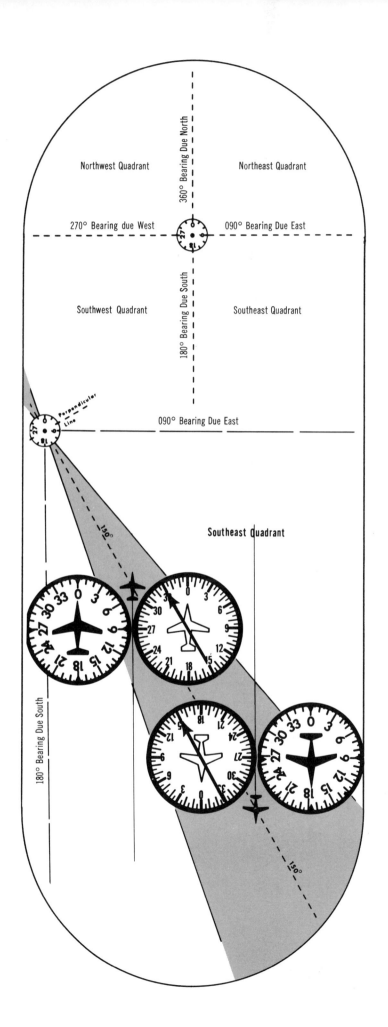

ORIENTATION AS TO QUADRANT

Another valuable tool of 'orientation' is to know quadrant (northeast, southeast, southwest or northwest) you are in while checking your course TO the NDB and the bearing/radial you are on. Knowing what quadrant you're in helps you 'visualize' your position in relation to the NDB. And it helps you in deciding what direction and heading you want to turn to.

The quadrants using the ADF Direction Indicator system of navigation are the same as the Course Deviation Indicator system of navigation. The Northeast quadrant encompasses all bearings/radials from 0° (360°) to the 090° bearing/radial. The southeast quadrant covers all bearing/radials from 090° to 180°. The southwest quadrant runs from 180° to 270°. The northwest quadrant then completes the circle by covering the area from 270° to 0° (360°).

HOW DOES KNOWING WHAT QUADRANT I'M IN HELP ME TO ORIENT MYSELF IN RELATION TO THE NDB TO WHICH I'M TUNED?

Looking at the accompanying illustration we see that your course TO the NDB is 330° (northwest) and you are on the 150° bearing/radial in the southeast quadrant. By knowing you are on the 150° bearing/radial you know you are in the southeast quadrant — the area between the 090° bearing/radial and the 180° bearing/radial, therefore you are southeast of the NDB. To help you visualize your position in relation to the (RBn) (NDB) (station) you have the Heading Indicator. Visualize yourself in the southeast quadrant then look at your Heading Indicator to determine the 'magnetic direction' in which you are flying. In this case it is 0° (360°). The course TO the NDB is 330° (northwest) therefore to go TO the NDB you would turn left to a heading of 330°.

MENTALLY OR PHYSICALLY POINTING IN THE DIRECTION OF THE NDB ALWAYS HELPS YOU VISUALIZE ITS POSITION IN RELATION TO YOU.

In the second illustration you are on the 150° bearing/radial again, but your heading has now changed to 180° (south); therefore to go TO the NDB you would turn **right** to a heading of 330°.

NOTE: We will talk about the angle of intercept later in this manual.

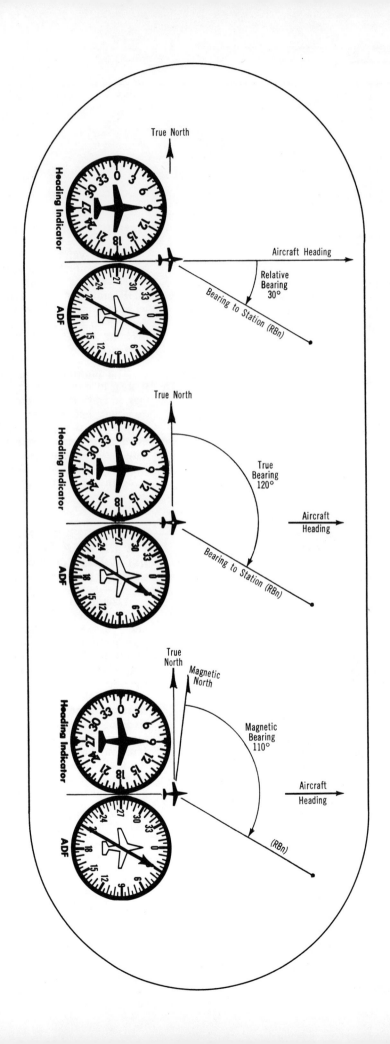

ADF DIRECTION INDICATOR BEARINGS

Of all the things concerning the ADF Direction Indicator system of navigation it is the 'bearings' that will confuse you the most. The word 'bearing,' as you may already know, is the terminology used by the men who sail the oceans and inland waterways. It is used to determine a position TO or FROM a navigation point on land or from ship to ship, aircraft to aircraft or aircraft to ship.

There is the 'true bearing' (measured clockwise from true north), 'magnetic bearing' (measured clockwise from magnetic north) and the 'relative bearing' (measured clockwise from the nose of your aircraft). ALL BEARINGS ARE ALWAYS MEASURED CLOCKWISE, either from true north, magnetic north or from the nose of your aircraft to the NDB to which you are tuned.

The illustration to your left shows how each bearing is arrived at using the formulas in the above paragraph. Study all three so you'll know when you're asked on the FAA exam. For study and illustration purposes in this manual we will concern ourselves with only the 'relative bearing,' i.e., the bearing/radial TO or FROM the NDB relative to the flight path of your aircraft. AGAIN, THIS IS ALWAYS MEASURED CLOCKWISE — **ALWAYS CLOCKWISE** — FROM THE NOSE OF YOUR AIRCRAFT.

The ADF Direction Indicator is 'tied in' to the NDB on the ground just as the Course Deviation Indicator in the Course Deviation Indicator system of navigation is 'tied in' to the Omni station on the ground. The Omni station on the ground emanates electronic signals to all points of the compass which are picked up by the Course Deviation Indicator in your aircraft telling you 'where you are' in relation to the station, i.e., what radial you are on. The NDB of the ADF Direction Indicator system of navigation also emanates electronic signals to all points of the compass. BUT we're not supposed to say that because technically it doesn't work that way, but it does emanate or radiate something because when you tune your ADF receiver to it the needle of your ADF Direction Indicator will point right to the NDB no matter where you are in relation to the NDB. So to make it less confusing, think of the NDB as having radials just like the OMNI and when you're going TO or FROM the NDB on a certain bearing think of it as a radial too!

To understand what bearing/radial you're on while using the ADF Direction Indicator system of navigation requires you to do mathematical calculations in your head in order to know what magnetic heading you should use to take you TO or FROM the NDB. You can do it by superimposing the ADF Direction Indicator needle onto the face of the Heading Indicator, but most all instructors and the FAA want you to be able to tell them how you arrive at the correct magnetic heading mathematically so let's study that for a moment. We will also be studying it as we progress through other illustrations in this manual.

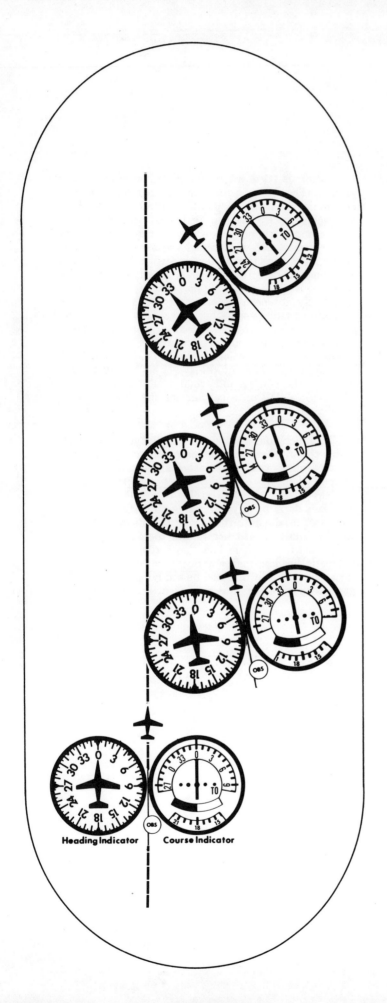

HOMING-OMNI

In ADF jargon it is also called bird-dogging. To accomplish 'homing/bird-dogging' using the Course Deviation Indicator system of navigation you must constantly keep centering the Course Deviation Indicator needle in order to keep 'on course' TO the station. I use the term 'on course' rather loosely here, because all you're really doing is flying TO the Omni, keeping the Course Deviation Indicator needle centered. So you're not really on a prescribed course, other than that prescribed by the wind. And that's what we mean when we say we're 'on course' in this particular instance.

Homing/bird-dogging is flying your aircraft TO an Omni regardless of your prescribed course (also referred to as your flight path) and is usually always a result of inattention to wind conditions affecting your prescribed course or flight path.

The illustration to your left shows a wind from your left blowing you off course and away from where you should be.

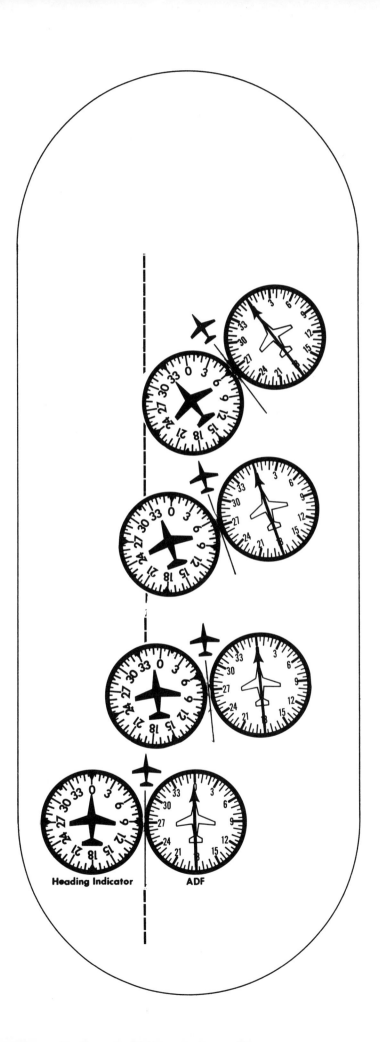

HOMING-ADF

Some old pilots call this 'bird-dogging,' i.e., always pointing. To accomplish homing/bird-dogging using the ADF Direction Indicator system of navigation you just keep the ADF Direction Indicator needle pointing to 0 (the nose of your aircraft) no matter what your flight path is over the ground. In the Course Deviation Indicator system of navigation where you must constantly keep centering the Course Deviation Indicator needle by constant adjustment of the OBS knob (Omni Bearing Selector knob). The ADF Direction Indicator system of navigation in contrast requires you to make adjustments with the ailerons and rudders on your aircraft in order to keep the needle centered on 0 (the nose of your aircraft).

As with the Course Deviation Indicator system of navigation 'homing/bird-dogging' results from inattention to wind conditions affecting your aircraft's flight path. And although you can get TO the station by 'homing/bird-dogging' using the Course Deviation Indicator or the ADF Direction Indicator both are ineffective ways of navigation.

Both were presented here to show what can happen in wind conditions when you are not attentive to your Heading Indicator and the Course Deviation Indicator and/or the ADF Direction Indicator.

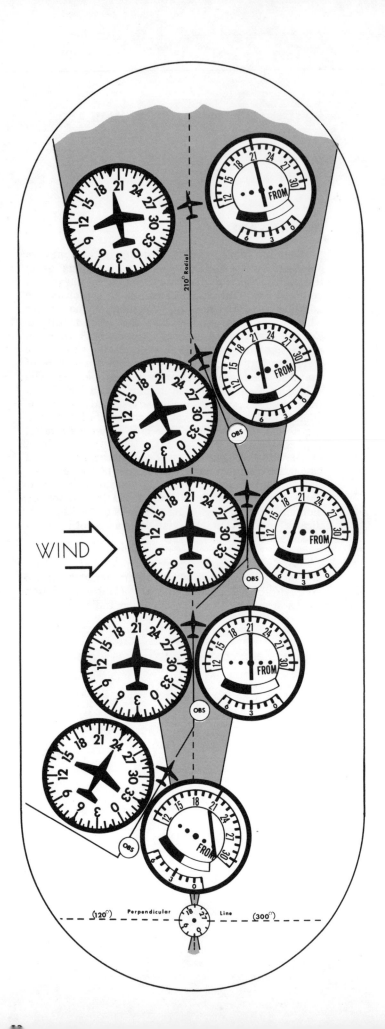

TRACKING OMNI

Keeping the CDI needle centered when there is a light-to-moderate or a moderate-to-heavy wind while going TO or FROM the Omni is known as 'tracking.'

When you're 'tracking' TO or FROM the Omni the CDI needle is the most important factor in helping you keep 'on course.' The Heading Indicator is important in that it tells you what heading to maintain in order to keep on a particular course. The Heading Indicator serves another useful purpose in that it tells you the difference, in degrees, between your 'heading' and the 'course' TO or FROM the Omni. This information is useful because it helps you in making the proper wind correction angle when you make a 'course' change.

The following method is probably the best for keeping you 'on course.' It minimizes the number of corrections you have to make to keep 'on course.' It requires the least amount of your attention. And it insures that you will pass over the 'top of the Omni' every time.

You intercept the 'course' as always making sure also that you maintain your heading. It is important to cross-check the Heading Indicator with the CDI needle of the Course Indicator because if the **CDI needle** starts to move 'off-center' you'll want to know why. More than likely **if it does not stay centered and you're heading is the same, the wind is blowing you 'off-course.'** If your CDI **needles** moves **left** you know that the **'course'** is to your **left** and if the CDI needle is **left** the **wind** is **from your left.** This means of course that you are to the right of the 'course.'

In the illustration you correct this by making a left turn of 20° to a heading of 190°. **Visualize** this by looking at the Heading Indicator i.e., the **'course'** is to your **left,** therefore to get back 'on course' would require you to turn left. Looking 20° to the left of your present heading reveals that to get back to the 'course' would require you to turn to a heading of 190°. You can subtract 20° from 210°, but it's easier and far less complicated if you will **visualize** your relationship to the 'course' and 'see' your new heading on the Heading Indicator. **THE REASON FOR THE 20° TURN IS THAT IN MOST WIND CONDITIONS 20° WILL RETURN YOU TO THE 'COURSE' AND IT IS SELDOM NECESSARY TO TURN MORE THAN 20° TO CORRECT FOR WIND DRIFT.**

Now, if you find that the new heading does not keep you 'on-course' i.e., if you over-shoot it, return to a heading half-way in between which in this case would be 200° and make whatever small corrections you need to stay 'on course' (CDI needle centered).

If the wind is very strong and the new heading does not take you back to the 'course' turn 10 more degrees left to a heading of 180° and then make your small corrections to remain 'on course' from there.

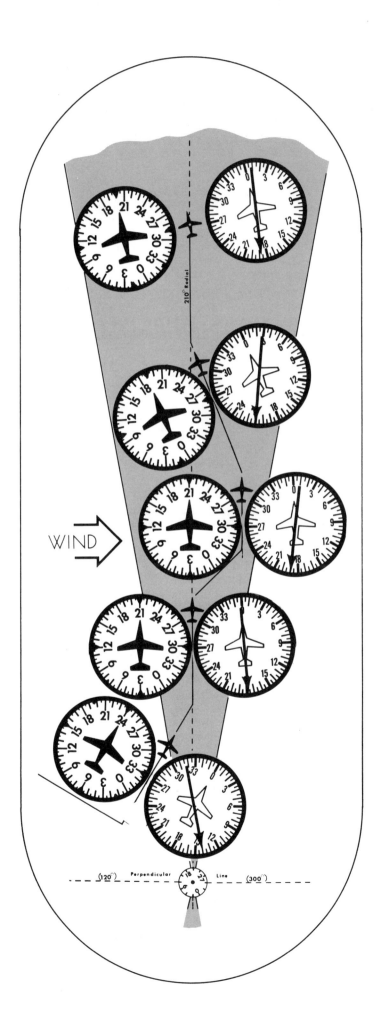

TRACKING ADF DIRECTION INDICATOR

Tracking, using the ADF Direction Indicator system of navigation, requires more concentration on your part than does the Course Deviation Indicator system of navigation because of the nature of the ADF Direction Indicator needle. Remember — IT ALWAYS POINTS TO THE NDB TO WHICH YOU ARE TUNED, NO MATTER WHERE YOU ARE IN RELATION TO THE NDB.

In the illustration to your left it shows you 'on course' after omtercepting the 210° bearing/radial outbound FROM the NDB. At #3 the wind from your left is pushing you 'off course' (to the right of your intended course or flight path) and by looking at the needle of your ADF Direction Indicator you note that the station is over your left shoulder in the area between the tail (18) of your aircraft and your left wing tip (27). Physically pointing in that direction will help you visualize your position in relation to the NDB. And again, THE NUMBERS ON THE ADF DIRECTION INDICATOR MEAN NOTHING AS FAR AS MAGNETIC DIRECTION IS CONCERNED. CONCERN FIRST WITH THE AREA IN WHICH THE NEEDLE IS POINTING, i.e., RIGHT OR LEFT OF THE NOSE AND RIGHT OR LEFT OF THE TAIL. When you are determining your 'relative bearing' then you can use the numbers but only then AS TO THE DEGREES MEASURED CLOCKWISE FROM THE NOSE OF YOUR AIRCRAFT, NOT MAGNETIC DEGREES.

The needle has moved to the left and to get back 'on course' you turn left. DON'T LOOK AT THE ADF DIRECTION INDICATOR NEEDLE WHEN YOU ARE TURNING—THAT WILL ONLY CONFUSE YOU. Turn to an intercept heading of at least 20° from your present heading. As with the Course Deviation Indicator system of navigation the 20° turn into the wind, in most wind conditions, is all that is required to return you to your 'desired course.'

WHEN THE ADF DIRECTION INDICATOR NEEDLE READS THE NUMBER OF DEGREES OF YOUR INTERCEPT ANGLE YOU ARE 'ON COURSE' AGAIN. In illustration #4 you turned 20° left from a heading of 210° to 190°, the ADF Direction Indicator needle is pointing to 21 (210°) but remember the 210° **here** has no magnetic significance as to where the NDB is; it only means at the moment that the NDB is left of your aircraft's tail. This particular aspect will probably confuse you more than anything because you'll always be wanting to attach some magnetic significance to the numbers — DON'T DO IT. The numbers are only useful to get the 'relative bearing' TO or FROM the NDB. Your ADF Direction Indicator needle, as you drifted 'off course' was pointing to 19 (190°). Adding 20° to 190° will give you 210° (21 on the dial). When your ADF Direction Indicator needle points to 19 (190°) or a least the mark on the dial that represents 19 you are back 'on course' and you should remain on that heading to make sure your ADF Direction Indicator needle holds at 19. Otherwise you'll be required to make another wind correction maneuver to keep you 'on course.'

27

TIME AND DISTANCE CHECKS

As you know, from the previous 'in-flight' exercises, you can find out approximately 'where you are' in relation to the Omni by utilizing the Course Indicator and the Heading Indicator. But you don't necessarily know how far away you are or how long it will take you to get there so let's discuss the problem of TIME AND DISTANCE CHECKS first.

THE ACCURACY OF TIME AND DISTANCE CHECKS ARE GOVERNED BY THE EXISTING WIND, THE DEGREE OF RADIAL CHANGE AND THE ACCURACY OF TIMING. The number of variables involved cause the result to be an approximation. However, **BY FLYING AN ACCURATE HEADING AND CHECKING THE TIME, AND RADIAL CHANGE CLOSELY YOU CAN GET A REASONABLE ESTIMATE OF THE TIME REQUIRED TO GET 'TO' THE OMNI and the DISTANCE IN MILES YOU ARE FROM THE OMNI.**

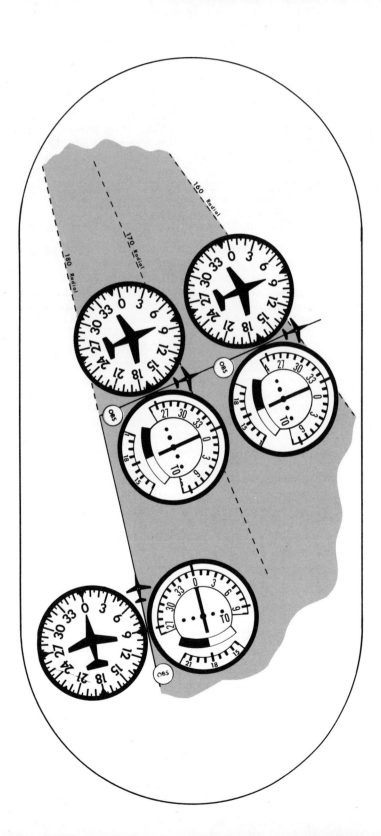

To compute time or distance TO the Omni you have to determine what heading you must turn to in order to cross the nearest radial at a 90° angle (the 90° angle is important because the formula is based on a right triangle theorem in geometry which says that the time to span the distance between two radials is proportional to the time needed to reach the Omni).

The **TIME REQUIRED TO GET 'TO' THE OMNI** is easily calculated when you **FLY A 10° RADIAL CHANGE and NOTE THE AMOUNT OF TIME (in seconds) REQUIRED TO FLY THAT 10° RADIAL CHANGE.**

The formula is written thus:

$$\frac{\text{time (in seconds) required to fly the 10° radial change}}{\text{10° radial change}} \text{ equals } \begin{array}{c}\text{time required}\\ \text{(in minutes)}\\ \text{to get 'to'}\\ \text{the Omni}\end{array}$$

Example:

$$\frac{75 \text{ seconds}}{10} \text{ equals } 7.5 \text{ minutes}$$

From the Course Indicator and the Heading Indicator indications in the illustration you know you are somewhere on the 180° radial and the 'selected course' is taking you TO the Omni. But you don't know how long it will take you to get there or how far away it is. You note that to cross the 170° radial at a 90° angle requires a heading change of 80° (170° — 90° = 80°). A change of 80° from your present heading of 0 (360°, due-North) would make your new heading 080°.

BEFORE YOU MAKE YOUR RIGHT TURN to 080° reset the Course Indicator to 350° (you still want a TO reading). Make your right turn to 080° and watch for the CDI needle to center. When it does note the time. Reset the Course Indicator to 340° (160° radial) and note the time when the CDI needle centers. Put that information into the above formula and you'll know how long it will take you to get TO the Omni.

To compute **THE DISTANCE YOU ARE FROM THE OMNI** you fly the same 10° radial change, note the time and put it in this formula:

$$\frac{\text{ground speed or true airspeed} \times \begin{array}{c}\text{time required,}\\ \text{(in minutes), to}\\ \text{fly the 10° change}\end{array}}{10°}$$

Example:

$$\frac{120 \text{ (ground speed)} \times \tfrac{3}{4} \text{ minutes}}{10°} = 9 \text{ miles}$$

YOU ARE 9 NAUTICAL MILES FROM THE OMNI.

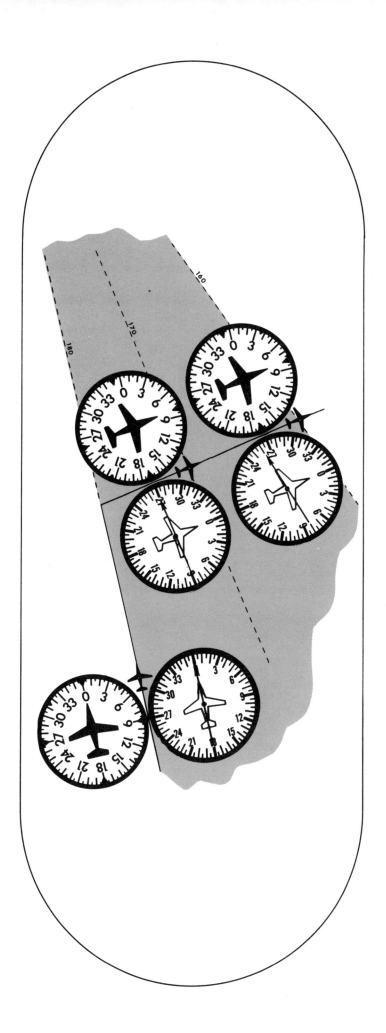

TIME AND DISTANCE CHECKS

As you know from the previous 'in-flight' orientation exercises you can find out approximately 'where you are' in relation to the NDB by utilizing the ADF and the Heading Indicator. But you don't necessarily know how far away you are or how long it will take you to get there, so let's discuss the problem of TIME AND DISTANCE CHECKS first.

THE ACCURACY OF TIME AND DISTANCE CHECKS ARE GOVERNED BY THE EXISTING WIND, THE DEGREE OF RADIAL CHANGE, THE ACCURACY OF TIMING. The number of variables involved cause the result to be an approximation. However, BY FLYING AN ACCURATE HEADING AND CHECKING THE TIME AND BEARING CHANGE CLOSELY YOU CAN GET A REASONABLE ESTIMATE OF THE TIME REQUIRED TO GET TO THE RBn AND THE DISTANCE IN MILES YOU ARE FROM THE RBn.

To compute time or distance TO the NDB you have to determine what heading you must turn to in order to cross the nearest bearing (radial) at a 90° angle. (The 90° angle is important because the time/distance formula is based on a right triangle theorem in geometry which says that 'the time to span the distance between two bearings (radials) is proportional to the time needed to reach the NDB.)

The time required to get TO the NDB is easily calculated when you fly a 10° bearing change and note the amount of time (in seconds) required to fly that 10° bearing change. The formula is written thus:

$$\frac{\text{time (in seconds) required to fly the 10° bearing change}}{10° \text{ bearing change}} = \begin{array}{l}\text{time required}\\ \text{(in minutes}\\ \text{to get TO}\\ \text{the NDB}\end{array}$$

Example: $\frac{75 \text{ seconds}}{10°} = 7.5$ minutes

From the ADF and Heading Indicator indications in the illustration you know you are somewhere on the 180° bearing (180° radial) FROM the NDB and that course is taking you TO the NDB. But you don't know how far away it is. You note that to cross the 170° bearing (170° radial) at a 90° angle requires a heading change of 80° (170—90 = 80). A change of 80° from your present heading of 0° (360° due North) would make your new heading 80°.

Make your right turn to 080°. When the ADF needle points to 27° you will know that you are on the 170° bearing (170° radial). Start timing. Stop timing when the ADF needle points to 260° (that's your 10° bearing [radial] change). Put that information into the above formula and you'll know how long it will take you to get to the NDB.

To compute the distance you are FROM the NDB you fly the same 10° bearing (radial) change noting the time and putting it into this formula.

$$\frac{\text{Ground speed/true airspeed}}{10°} \times \begin{array}{l}\text{time required}\\ \text{(in minutes)}\\ \text{to fly the 10°}\\ \underline{\text{change}}\end{array}$$

Example: $\frac{120 \text{ (ground speed)} \times \frac{3}{4} \text{ mins.}}{10°} = 9$ miles

You are 9 nautical miles from the NDB.

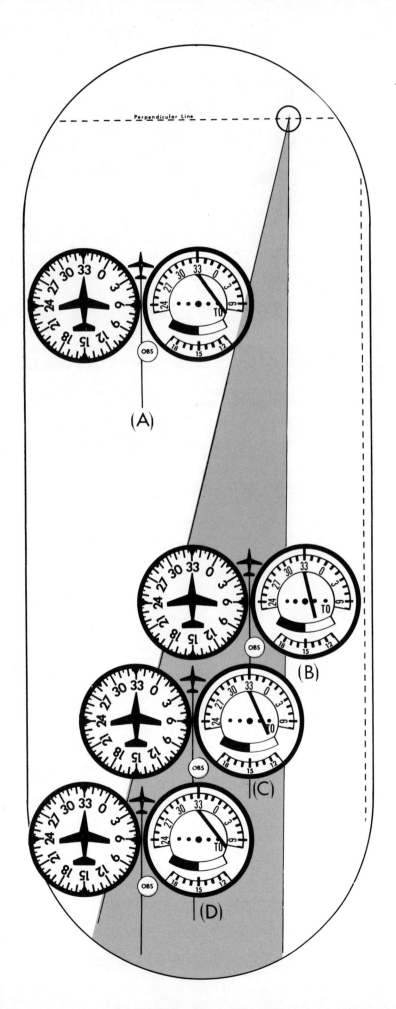

INSIDE AND OUTSIDE THE CDI NEEDLE SENSITIVITY AREA

You've oriented yourself in relation to the Omni and you've found that you're approximately 9 miles from the Omni it will take you approximately 4½ minutes to get to the Omni.

You've decided to go TO the Omni on the 150° radial on a 'course' of 330°. Once you decide the 'course' TO the Omni one of two things will occur. One, you may be **outside** the CDI needle sensitivity area as at A, or two, you may be **inside** the CDI needle sensitivity area (shaded portion) as at B, C. and D.

If you are **outside** the CDI needle sensitivity area the CDI needle will be 'pegged' (right or left whatever the case may be — in this particular case it is right).

If you are **inside** the CDI needle sensitivity area the CDI needle will sense approximately how close you are to the 'selected' course as set-up on your Course Indicator and you can use this information to help you decide the heading you should fly in order to intercept your 'selected' course at the proper intercept angle.

If you are **outside** the CDI needle sensitivity area you parallel the 'selected' course, note your heading then intercept the 'selected' course at no less than a 45° angle. In this case your intercept heading would be 015° (045°+330°=375°—360°=015°). Knowing that you're going to intercept the 'selected' course at a 45° angle simply look at the Heading Indicator and note the reading under the first mark 45° to the right of the center large mark and turn to that heading. That way no mental calculations are needed.

If you are **inside** the CDI needle sensitivity area as at B, C and D you can, by paralleling the 'selected' course, determine the best intercept angle by noting the location of the CDI needle.

At B the needle is deflected a small amount so you're close to the 'selected' course and in this case an intercept angle of 20° would be sufficient. Your intercept heading would then be 350°.

At C the CDI needle is deflected a little more than at A so you're not as close to the 'selected' course. A sufficient intercept angle here would be 30°. The intercept heading would be 360° (due-North).

At D the CDI needle is almost 'pegged' to the right. You're almost on the 10° line so an intercept angle of 45° would be the best here. Your intercept heading would be 015°.

These illustrations are not to scale. They are drawn this way to give you some idea of what to look for.

Note too, that it is best to make sure you fly and understand the Orientation Exercises before you work on the Interception Exercises.

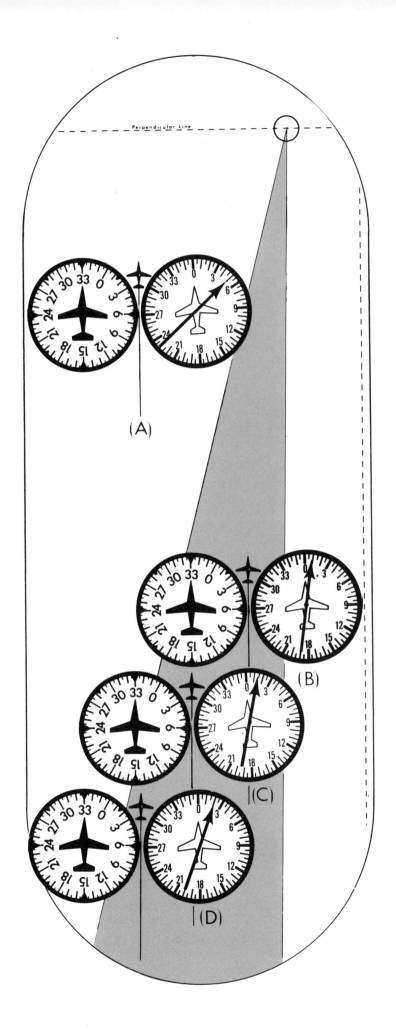

AREA OF ADF DIRECTION INDICATOR NEEDLE SENSITIVITY

That's a misnomer of course, because the needle of the ADF Direction Indicator is always sensitive, i.e., it always points TO the NDB no matter where you are in relation to the NDB, but I left the illustration there to show the similarity of aircraft on the same flight path as they were in the NDB Course Deviation Indicator illustration in relation to the NDB to which they are tuned. The three aircraft at D, C and B are within what would be called the Course Deviation Indicator sensitivity area and only D's ADF Direction Indicator needle would be close to 10° to the right of the nose of your aircraft (the shaded area in the illustration). And if you added that 10° to your Heading Indicator's reading of 33° (330° + 010°) you would have a 'relative bearing' of 340°. That's not precisely true in all three cases because they are all different 'relative bearings.'

Illustration (A) is outside the 10° area and its 'relative bearing' TO the NDB would be 330° + 040° (the ADF Direction Indicator needle is pointing to the mark that represents 40°) = 370° and since the compass only goes to 360° you must subtract 360° from 370° (10°) to get your 'relative bearing' TO the NDB. Turning to a magnetic heading of 040° on your Heading Indicator will take you TO the NDB. DON'T WATCH THE ADF DIRECTION INDICATOR AS YOU TURN — JUST WATCH YOUR HEADING INDICATOR SO YOU CAN 'ROLL OUT' ON THE PROPER HEADING. If your calculations are correct your ADF Direction Indicator needle will point to (0°) the nose of your aircraft. Remember the NDB will be right in front of you. AFTER YOU ARE ESTABLISHED ON YOUR DESIRED HEADING **IN WINGS LEVEL FLIGHT.** The foregoing is very important — DON'T WATCH THE ADF DIRECTION INDICATOR NEEDLE WHEN TURNING AND DON'T LOOK AT IT AGAIN UNTIL YOU ARE IN **WINGS LEVEL FLIGHT.** Look at it **then** to assess your situation as to where the NDB is **in relation to the nose of your aircraft.**

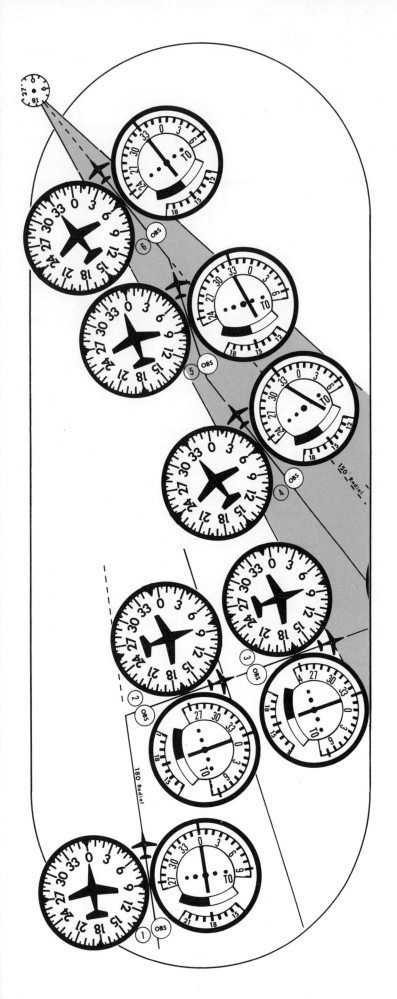

(Before reading this section page ＿-page ＿, we suggest you unfold the foldout page 49 and leave extended for study and reference purposes.)

OMNI INTERCEPTION

I have tried in all cases to make the illustrations, and the written text material, pertinent to the 'in-flight' exercises and as close as possible to real-life flight situations. Each of the following pages are excerpts from the Flight Path Sequence Study Sheet and it is the purpose of each to illustrate a particular phase of interception and show you how that particular phase should be accomplished. The illustrations are shown in a 'no-wind' condition, i.e., the air is calm, so if your heading while flying is different from that illustrated it's because you're keeping the CDI needle centered and **THAT'S EXACTLY WHAT YOU SHOULD ALWAYS DO.**

Fold out the Flight Path Sequence Study Sheet page 49.

INTERCEPTION OF THE 330° RADIAL 'INBOUND' TO THE OMNI (STATION)

After you note the time required to make the 10° radial change at #3 you reset your Course Indicator to 33 (330°, Northwest), so you can intercept the 150° radial and go TO the Omni, you will note that in all probability that the CDI needle will swing to the **LEFT INSTEAD OF RIGHT. IF IT DOES IT IS BECAUSE AT THE PRESENT TIME YOU ARE ON AN INTERCEPT HEADING OF THE 150° RADIAL AT AN ANGLE OF MORE THAN 90°. KNOW THAT THE CDI NEEDLE WILL ALWAYS REVERSE WHEN THE ANGLE OF INTERCEPT IS GREATER THAN 90°, THEREFORE THAT IS THE REASON YOU SHOULD NEVER INTERCEPT A RADIAL AT AN ANGLE GREATER THAN 90°.**

So, in this particular case you made a left turn to 330° so you would **PARALLEL THE 'IN-BOUND' RADIAL** (#4) to determine what angle (and therefore what heading) would be required to intercept the 150° radial. In the illustration you can see that a 20° angle would be sufficient, however when you fly the 'in-flight' exercises another angle may be more suitable. YOU decide that when you're there.

At #5 you changed you're heading in order to intercept the 150° radial at the required angle.

And at #6 you're 'in-bound' TO the Omni on the 150° radial.

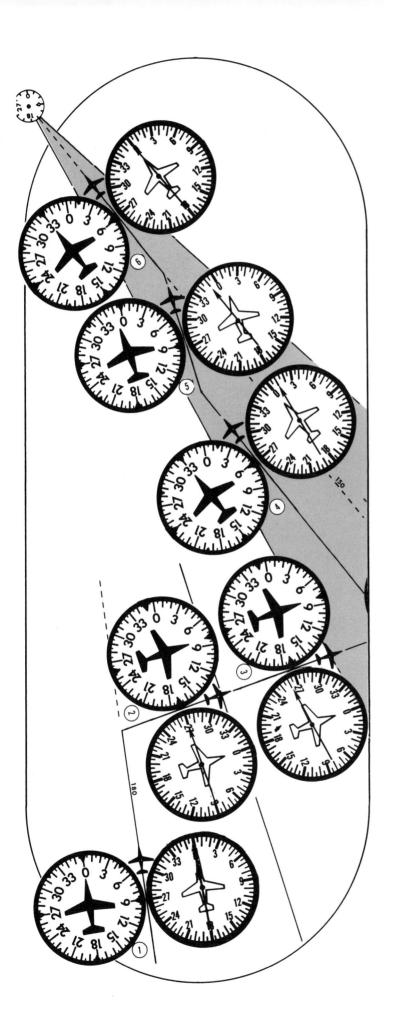

NDB COURSE INTERCEPTION

Fold out the Flight Path Sequence Study Sheet on page 49 so you will have some reference point as you study each Course Deviation Indicator and ADF Direction Indicator illustration and written text sequence on the following pages.

Study both the Omni and NDB illustrations as you go through this manual so you'll be able to see how similar they are.

INTERCEPTION OF THE 330° BEARING/ RADIAL 'INBOUND' TO THE NDB

You'll notice that positions #1, 2 and 3 are part of the Time and Distance illustration on page 29 and are presented here for you perusal. Number one (1) is also the beginning position on the Flight Path Sequence Study Sheet.

At position #2 you will note that the station is directly off your left wing tip (27). Look or point in that direction so you'll orient yourself as to its location. By superimposing the ADF Direction Indicator needle onto the Heading Indicator dial the course TO the station at this particular point is 350°. You're within 20° of your course 'inbound' TO the station. Mathematically figuring your 'relative bearing' TO the station you have 080° (Heading Indicator) + 270° (ADF Direction Indicator reading) = 350° (your 'relative bearing' TO the NDB at this particular point.

As with the Course Deviation Indicator system of navigation WHEN YOU ARE ON AN INTERCEPT HEADING OF MORE THAN 90° (as you are here at #3) you make a left turn (DON'T WATCH THE ADF DIRECTION INDICATOR NEEDLE AS YOU'RE TURNING) to 330° (33 on your Heading Indicator) to parallel the 330° bearing/radial 'inbound TO the NDB.

Quickly establish yourself on a heading of 330° then make a 20° turn to 350° in intercept the 330° bearing/ radial TO the NDB. WHEN THE ADF DIRECTION INDICATOR NEEDLE EQUALS THE NUMBER OF DEGREES OF YOUR INTERCEPT ANGLE (here 350°) YOUR 'ON COURSE' and you now turn your aircraft to a heading of 330°. At this point, all things being equal, the ADF Direction Indicator needle will point to 0 — the nose of your aircraft.

NOTE: Before you made the 20° turn to intercept the 330° bearing/radial the ADF Direction Indicator needle pointed to 10 (or at least the mark that represents 10). You **don't add** 20°, you **subtract it.** When you're in your aircraft this can be very confusing, so that's why it's very important to visualize 'where you are' in relation to the NDB (station). **At #5 the ADF Direction Indicator needle setting of 350° has nothing to do with your aircraft's magnetic heading, but it is equal to the 20° of your intercept angle which means that you're 'on course.'**

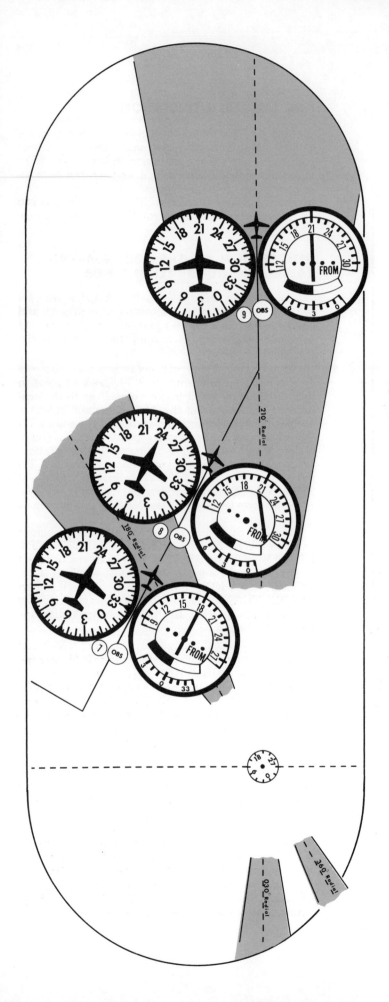

OUTBOUND INTERCEPTION OUTSIDE THE CDI NEEDLE SENSITIVITY AREA

Now let's consider what you would do if you were **outside** the CDI needle sensitivity area.

We'll continue on from position #6 in the preceding illustration and assume that you've decided to intercept the 210° radial and fly 'outbound' FROM the Omni. After making a 90° heading change you wanted to check your progress so you rotated the OBS knob so that 18 would show in the upper window (18 **BECAUSE YOU KNOW YOU'RE GOING TO GO 'OUTBOUND' FROM THE OMNI**) to see if you had passed the 180° radial. In the illustration before, you used a TO indication because you wanted to go TO the Omni. Now you want to go away FROM the Omni. Using the FROM indication for a radial passage is okay in this type of case. The new 'selected course' is 210° which means that once you find out when you cross the 180° radial you only have to rotate the OBS knob 30° to be set-up on your new 'selected course'—210° (outbound FROM the Omni).

As always, **VISUALIZATION IS IMPORTANT.** Your new heading when 'on course' will be 210°. Which means that you'll be going Southwest and you'll be in the Southwest Quadrant. **VISUALIZE THAT WHEN YOU FLY THIS PART.**

The Course Indicator tells you when you reach the 180° radial as in number 7 illustration. After you cross the 180° radial and the CDI needle starts to move rotate the OBS knob to your 'selected course' of 210°.

Your heading is 240°. Your 'selected' course is 210°. MENTAL CALCULATION tells you the intercept angle will be 30°, but get in the habit of looking at your Heading Indicator and VISUALIZING your relationship to the Omni and making calculations that way too.

When the CDI needle is almost centered as you approach the 210° radial start your turn to intercept the 210° radial. It takes practice to be able to intercept the 'course' at the right moment everytime, but with practice you'll do it easily.

When you reach position #8 watch how fast the CDI needle moves toward center. This will help you judge when you should start your turn to intercept your 'selected course.'

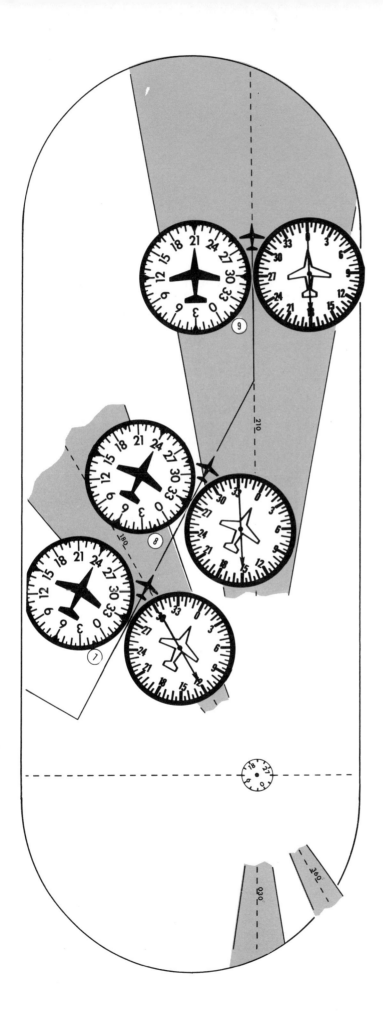

OUTBOUND NDB INTERCEPTION

Before reaching the NDB 'inbound' on the 150° bearing/radial you are instructed to intercept, and fly 'outbound,' on the 210° bearing/radial. You know 'where you are' now — 'inbound' on the 150° bearing/radial and in the southeast quadrant — and you've been instructed as to 'where to go' — 'outbound' on the 210° bearing/radial and you know that that bearing/radial is in the southwest quadrant. Because of your present direction of flight you know that the southwest quadrant is to your left — look that way to help you visualize your location as it relates to 'where you want to go' — so you know that a left turn will take you to your new 'course.' At position #6 your Heading Indicator reads 330° (33). A left turn to 210° will not intercept your new course, so you must first decide on a correct 'angle of intercept' so you'll be established 'on course' as quickly as possible. Thirty degrees in most cases. Using that as your basis and looking at your Heading Indicator you note that a heading of 240° would allow you to intercept your desired course at a 30° angle. FORGET ABOUT YOUR ADF DIRECTION INDICATOR WHILE TURNING. ESTABLISH YOURSELF ON A HEADING OF 240° THEN WATCH THE ADF DIRECTION INDICATOR NEEDLE MOVE FROM THE AREA JUST ABOVE YOUR RIGHT WING TIP TO BELOW YOUR RIGHT WING TIP. The ADF Direction Indicator needle always points to the NDB.

At position #7 the NDB is just behind your right wing tip so physically look in that direction to orient yourself and make a mental note that for me to be 'on course' the ADF Direction Indicator needle will be pointing to 18 on the dial — the tail of my aircraft. THE NUMBER ON THE ADF DIRECTION INDICATOR MEANS ABSOLUTELY NOTHING MAGNETICALLY AT THIS TIME.

What the numbers do mean is that when the needle of the ADF Direction Indicator points to the number that represents 30° to the right of your aircraft's tail (15) you are 'on course' and must make a left turn to a heading of 210° so that you remain 'on course.' Your ADF Direction Indicator is pointing at 18 — the tail of your aircraft.

REMEMBER — when the needle deflection of your ADF Direction Indicator from the right or left of the nose or tail of your aircraft equals the number of degrees of your angle of interception (here 30°) you are on your intended course, but you must make the turn to your desired heading (using the Heading Indicator) before you are actually 'on course.' The tail of your aircraft is 18, so 30° from the tail position equals 150° or (15).

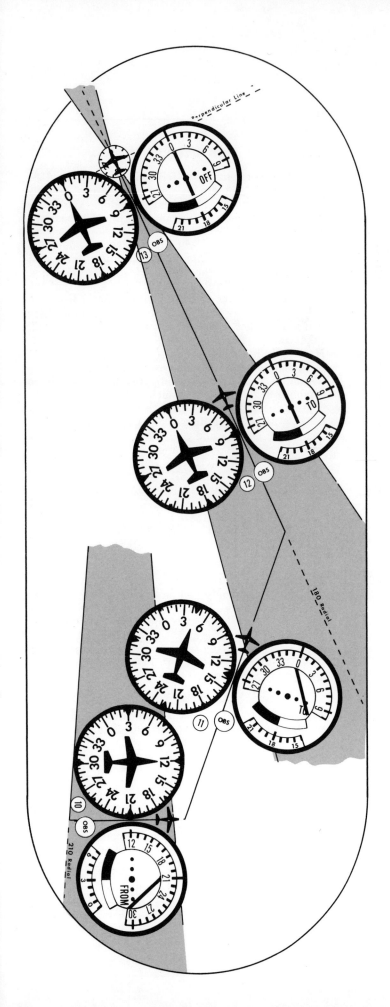

'INBOUND' OMNI INTERCEPTION 'OUTSIDE' THE CDI NEEDLE SENSITIVITY AREA

This illustration depicts another instance where the 'selected course' TO the Omni and your present heading will make the intercept angle greater than 90°. **YOU SHOULD NEVER INTERCEPT A RADIAL AT AN ANGLE GREATER THAN 90° BECAUSE OF THE VERY LIKELY POSSIBILITY THAT THE CDI NEEDLE WILL REVERSE AND MAKE YOU THINK THAT THE 'SELECTED COURSE' IS THE OPPOSITE OF WHERE IT REALLY IS.**

After traveling 'outbound' FROM the Omni, on the 210° radial you initiate a 90° left turn to a heading of 120° and decide at that time to go 'inbound' TO the Omni on the 180° radial. Looking at your Heading Indicator at position #10, you notice that on your present heading of 120° the angle of intercept will be greater than 90°. Therefore you decide to intercept the newly 'selected course' at an angle of 45°. Having made that decision you calculate that your intercept heading will have to be 045°. Looking at your Heading Indicator you can see that that is correct (45° to the right of 0).

At position #10 you are maintaining your present heading of 120° until the CDI needle is 'pegged' to the right indicating that you are out of the CDI needle sensitivity area. Then you make your left turn to a heading of 045° to intercept the 180° radial 'inbound' TO the Omni. After making the left turn to 045° reset the Course Indicator to 0. The CDI needle will be 'pegged' to the right and will start to move when you come within the 10° CDI needle sensitivity area again. It is at this time that you should keep a close watch on it to see how fast it moves toward the center of the dial. **THIS IS YOUR CLUE AS TO WHEN TO BEGIN YOUR TURN 'INBOUND' TO THE OMNI ON THE 180° RADIAL.**

At position #11 you'll see the first indication of the CDI needle start its move toward the center of the dial.

At position #12 you are 'on course' TO the Omni.

At position #13 you are 'on top' the Omni and the instrument indications read just like they did in the ORIENTATION exercises when you crossed the Omni (and the perpendicular line).

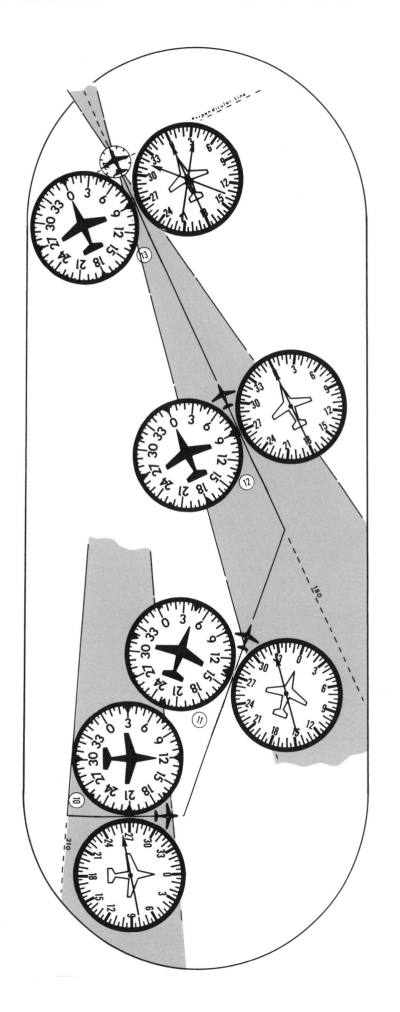

INBOUND NDB BEARING/RADIAL INTERCEPTION

After a few minutes of flying 'outbound' on the 210° bearing/radial you are instructed to turn left to intercept the 180° bearing/radial from your present position and fly 'inbound' TO the NDB.

You know that you are on the 210° bearing/radial 'outbound' FROM the NDB and that you are in the southwest quadrant. Looking at the Heading Indicator at position #9 you note that the 180° bearing/radial is to your left — physically look that way to orient yourself.

Even though you are to intercept the 180° bearing/radial your magnetic heading TO the NDB will be 0° (360°) not 180°. It is very important that you understand this and in looking at your Heading Indicator at position #9 it looks as though turning right to 0° would be the shortest and fastest route to the 180° bearing/radial. But you were instructed to turn LEFT to intercept the 180° bearing/radial so you must decide FIRST what magnetic direction you want to turn to. You decide on 90° so your heading at position #10 is 120° (90° from your heading 210°). You also note from looking at your Heading Indicator that your angle of interception of the 180° bearing/radial at this point is still greater than 90°. Rather than make a turn to parallel your 'inbound' course you decide on an angle of intercept of 45°. DON'T CONCERN YOURSELF WITH THE ADF DIRECTION INDICATOR READING AT THIS TIME OTHER THAN TO NOTE WHERE THE NDB IS IN RELATION TO YOUR AIRCRAFT AND AT POSITION #10 IT IS OFF YOUR LEFT WING TIP — LOOK IN THAT DIRECTION TO ORIENT YOURSELF. Mathematically speaking the 'relative bearing' TO the NDB is 30°, i.e., your magnetic heading (120°) plus your ADF Direction Indicator reading (270°) equals 390°. You now must subtract 360° from 390° and you get 30° — the 'relative bearing' TO the NDB at that point. You can also get it by superimposing the ADF Direction Indicator needle onto your Heading Indicator.

Now, since you decided on an angle of intercept of 45°, that 45° must be to the **right** of 0° (360°) — the heading that you must take up once you are on the 180° bearing/radial. So make another left turn and roll out on a heading of 045° and when you are in WINGS LEVEL FLIGHT check your ADF Direction Indicator needle so that you'll know that when the needle is 45° to the left of the nose of your aircraft you are 'on course' and must now make your left turn to 0° (360°) — position #11a.

At position #12 you are 'on course.'

At position #13 you are 'on top' the NDB and depending on whether you are a little to the right or left of the NDB the needle will swing that way, i.e., right or left of the top position.

37

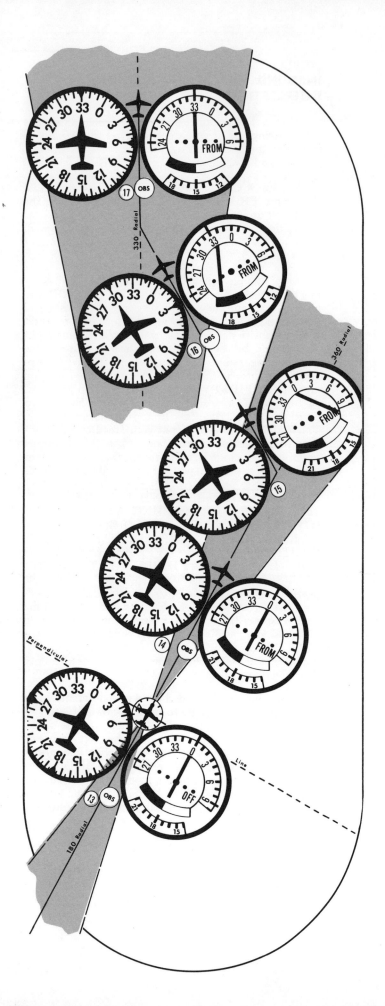

INTERCEPTING AN 'OUTBOUND' RADIAL AFTER CROSSING THE OMNI

Position #13 illustrated here is the same #13 as on page 36. It is presented here for continuity.

After flying 'outbound' on the 360° radial you decide to intercept the 330° radial, also flying 'outbound' FROM the Omni.

Your present heading at position #13 and 14 is 0 (360° due-North). Your new 'selected course' is 330° (Northwest). You want to intercept the new course at a 30° angle. Looking at your Heading Indicator at position #14 you note that in order to intercept the 330° radial with a 30° intercept angle you must turn **LEFT** to an intercept heading of 300°.

At position #15 you make the left turn and watch for the 'pegged' position of the CDI needle so that you'll know when you leave the CDI needle sensitivity area. When you do you can reset the Course Indicator to your new 'selected course' of 330° (33 on the Course Indicator dial) and when you enter the CDI needle sensitivity area, position #16, you'll notice the CDI needle start to move towards the center of the dial. Also note that when you reset the Course Indicator the CDI needle will swing left indicating that the 'selected course' is to your left.

DID YOU NOTICE THAT THE 'SELECTED COURSE' AND THE RADIAL WERE THE SAME?

When you reach position #16 notice how fast the CDI needle moves toward the center of the dial. This will help you judge when you should start your turn to intercept the 330° radial 'outbound' FROM the Omni.

At position #17 you are 'on course' 'outbound' FROM the Omni.

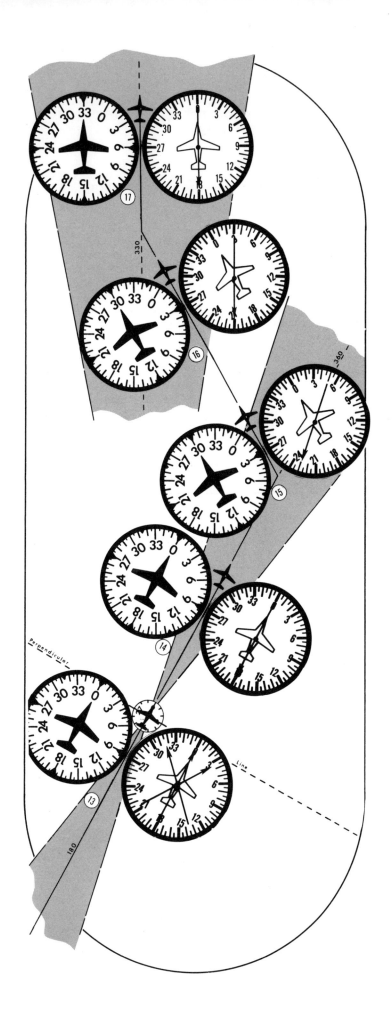

INTERCEPTING AN OUTBOUND BEARING/ RADIAL AFTER CROSSING THE NDB

At position #14 you have just crossed the NDB. The NDB is right behind you, i.e., the ADF Direction Indicator needle is pointing to 18 — the tail of your aircraft. ONLY BECAUSE YOU'RE ON THE 0° (360°) RADIAL DOES THE 0° AND 180° COINCIDE WITH THE MAGNETIC DIRECTIONS.

You are now instructed to turn left to intercept and fly 'outbound' on the 330° bearing/radial FROM the NDB.

Noting your position in relation to the NDB you see that your new course will be in the northwest quadrant and that the course is to your left — physically look in that direction to orient yourself.

Your next decision is — 'what angle of intercept?' Except in special occasions (your last intercept for one) a 30° angle of intercept is the most desirable. That being the case a 30° angle of interception will mean a left turn to a heading of 300°. DON'T WATCH THE ADF DIRECTION INDICATOR NEEDLE WHILE MAKING THIS (OR ANY OTHER) TURN.

Turn to a heading of 300° and AFTER ESTABLISHING YOURSELF IN WINGS LEVEL FLIGHT NOTE THE POSITION OF THE ADF DIRECTION INDICATOR NEEDLE. When it is within 30° of the tail of your aircraft (21) on the ADF Direction Indicator dial you are 'on course' and should turn right to your new heading of 330°.

Your angle of interception is 30° so your clue as to when you have reached the 330° bearing/radial is when the ADF Direction needle is 30° to the left of the tail of your aircraft, i.e., the 30° mark (21) is equal to the number of degrees (30°) of your interception angle.

You'll note that when you fly this particular flight path in your aircraft that if you are more than ten (10) minutes of flying time from the NDB you will have no appreciable error if you delay your turn onto your desired bearing/radial until the ADF Direction Indicator needles deflection equals the angle of interception. However, when you are less than ten (10) minutes flying time from the NDB your turning radius will cause you to overshoot if you wait too long to begin your turn.

39

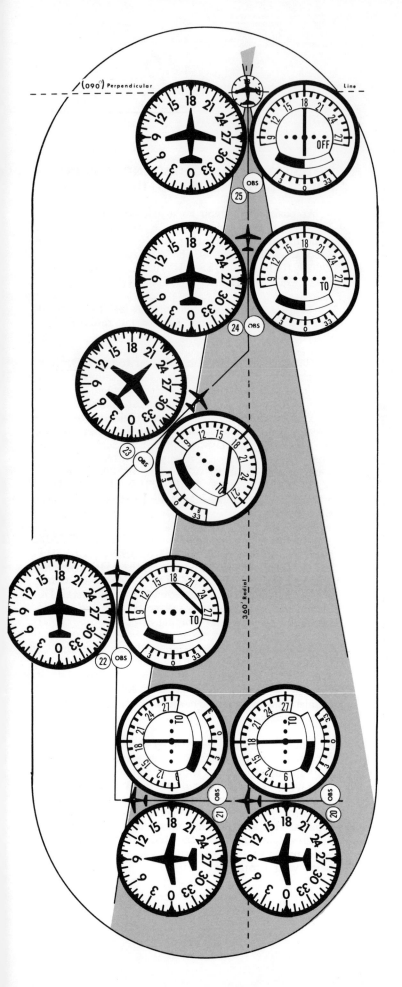

TIME AND DISTANCE CHECKS PLUS INTERCEPTING AN 'INBOUND' RADIAL FROM OUTSIDE THE CDI NEEDLE SENSITIVITY AREA

Your heading at position #20 is such that you'll be crossing the 360° radial at a 90° angle so let's do another **TIME AND DISTANCE CHECK** to see how far you are from the Omni and how long it will take you to get TO the Omni.

Before reaching position #20 reset the Course Indicator to 18 (180° due-South). You want a TO reading. You are on the TO side of the perpendicular line and the only way to get the TO reading **and** have the CDI needle center when you cross the 360° radial is to dial in 18 (180° due-South) on the Course Indicator.

As soon as the CDI needle centers begin timing and then reset the Course Indicator to 19 (10° to the right of 18) so that you'll know when the CDI needle centers again and therefore be able to calculate the **time** to insert in the **TIME AND DISTANCE** formula. Incidentally, if the CDI needle moves left you've inserted the wrong data into the Course Indicator.

After noting the time, turn right to a heading of 180° to parallel the newly 'selected course' of 180°, then reset the Course Indicator to 18 (180° due-South). If you've inserted the correct data the CDI needle will be 'pegged' to the right.

You'll be **OUTSIDE** the CDI needle sensitivity area so plan on an intercept angle of 45°. You'll be turning **RIGHT** to intercept your 'selected course' so look at the 45° mark on the **RIGHT** side of the Heading Indicator dial (position #22) and note the intercept heading you'll turn to, then make that turn to your intercept heading. The intercept heading in this case will be 225°.

Note the rate at which the CDI needle moves after you enter the CDI needle sensitivity area at position #23. Here again the CDI needle movement is your clue as to when to initiate your turn 'inbound' to get 'on course.'

At position #24 you're 'on course' 'inbound' TO the Omni.

At position #25 you're again 'on top' the Omni and at the same time crossing the perpendicular line, therefore your Course Indicator will read OFF.

$$\frac{\text{time (in seconds) required to fly the 10° radial change}}{\text{10° radial change}} \text{ equals } \begin{array}{c}\text{time required}\\\text{(in minutes)}\\\text{to get 'to'}\\\text{the Omni}\end{array}$$

$$\frac{\text{ground speed or true airspeed}}{10°} \times \begin{array}{c}\text{time required}\\\text{(in minutes), to}\\\text{fly the 10° change}\end{array} = \begin{array}{c}\text{miles}\\\text{from the}\\\text{Omni}\end{array}$$

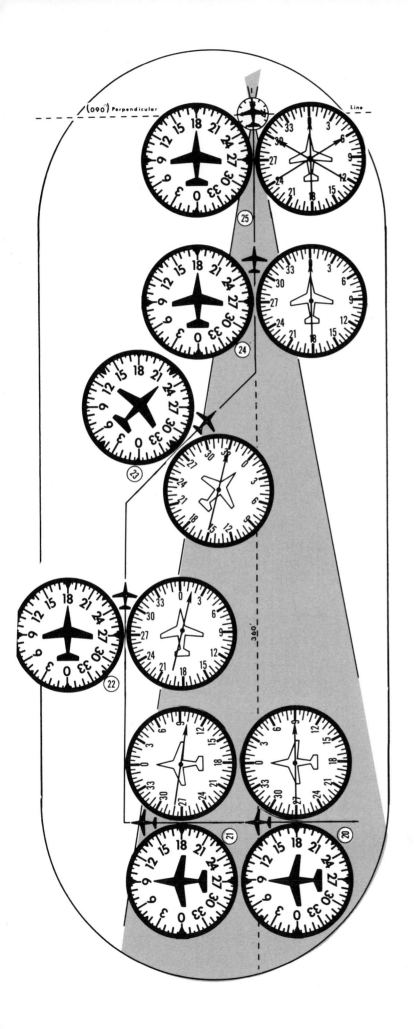

TIME & DISTANCE CHECK PLUS INTERCEPTING AN INBOUND BEARING/ RADIAL TO THE NDB

You're now instructed to make a Time and Distance check before intercepting the 360° bearing/radial 'inbound' TO the NDB.

At position #17 you are on the 330° bearing/radial 'outbound' FROM the NDB. In order to do the time and distance check **and** intercept the 360° bearing/radial you decide to turn right to a heading of 360° (0°) and make a right 90° turn to a heading of 090°. This will put you on a heading that will intercept the 360° bearing/radial at 90° so you can time the 10° bearing/radial change needed to compute the time and distance problem.

After making the turn to 090° keep your eye on the ADF Direction Indicator needle and when it reaches your right wing tip position (9) you know you are crossing the 360° bearing/radial. Start your timing now and finish it when the needle points to the mark that represents a 10° bearing/radial change and put it in this formula:

$$\frac{\text{time (in seconds) required to fly the 10° bearing/radial change}}{10° \text{ bearing/radial change}} = \text{time required (in minutes) to get to the NDB}$$

To make a right turn to 180° to parallel the desired course 'inbound' TO the NDB. After establishing your aircraft IN WINGS LEVEL FLIGHT note that the ADF Direction Indicator needle says the NDB (station) is to the right of the nose of your aircraft. THE NUMBER OF DEGREES DOESN'T MAKE ANY DIFFERENCE. Your only concern at the moment is the course you must fly to intercept the 360° bearing/radial. You decide at this time to try a 45° intercept angle. Looking at your Heading Indicator you note that a 45° intercept angle means a heading 225°.

Just after you make your turn to a heading 225° you will note that your ADF Direction Indicator needle moves **away** from the nose of your aircraft and that's what it should do. When it reaches a point where it is 45° to the left of the nose of your aircraft (here 315°) — or at least the mark that represents 315° — you are 'on course' and should make the left turn to 180°.

These illustrations and text are all shown in a no-wind condition and it is important that when you fly that the correct drift correction-angle be established as soon as possible after interception of an 'inbound' or 'outbound' bearing/radial. With the 'course' pinned down you will be more alert to signs of station approach.

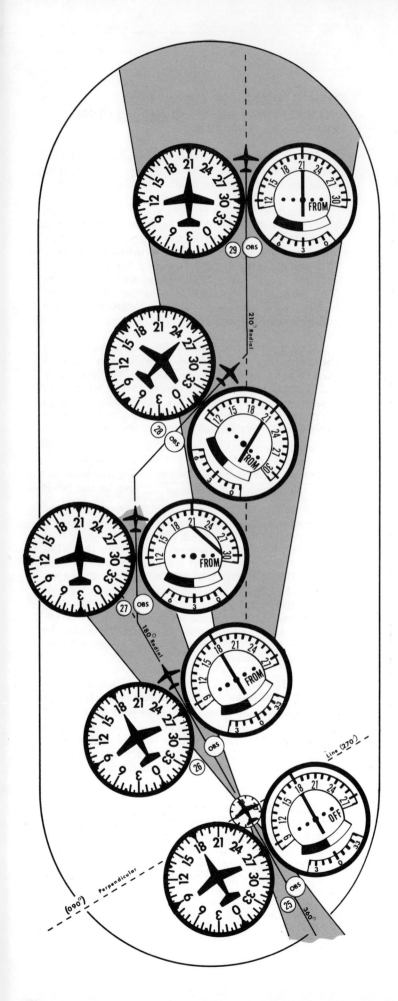

INTERCEPTION OF AN 'OUTBOUND' RADIAL IMMEDIATELY AFTER CROSSING THE OMNI

Position #25 is the same #25 from the preceeding page. It is presented here for continuity.

Wait until you get a firm FROM reading (position #26) before you start your RIGHT turn to parallel the newly 'selected course' of 210°, Southwest (21 on your Course Indicator dial).

After you parallel the 'selected course' reset your Course Indicator to 21 (210° Southwest), position #27. The 'selected course' is to your RIGHT so the CDI needle will be 'pegged' RIGHT, because you are OUTSIDE the CDI needle sensitivity area.

Since you're outside the CDI needle sensitivity area intercept the 'selected course' with a 45° intercept angle. The Heading Indicator at position #27 reveals that the intercept heading will be 255° (45° RIGHT of your present heading).

At position #28 the CDI needle is moving towards the center of the dial. Its speed will be your clue as to when to start your turn to the 'selected course' of 210°.

At position #29 you're 'on course' 'outbound' FROM the Omni.

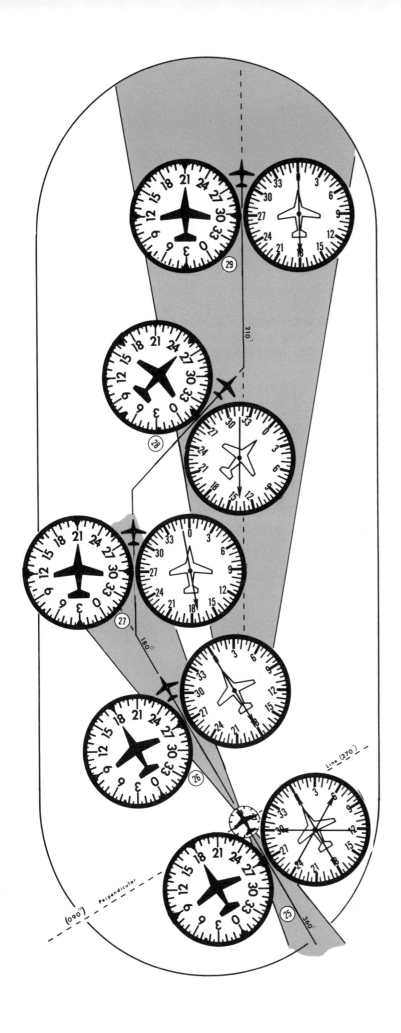

INTERCEPTION OF AN OUTBOUND BEARING/RADIAL IMMEDIATELY AFTER CROSSING THE NDB

Beware, as at position #25, that when you get close to the NDB that the ADF direction needle will in all probability start to swing pretty wildly. When it does you know you're close to the NDB.

Your instructions now are to track 'outbound' on the 210° bearing/radial FROM the NDB.

At position #26 the NDB is behind you — note where the ADF Direction Indicator needle is pointing — and you are 'outbound' FROM the NDB.

Concentrating on your Heading Indicator reading and your knowledge of 'where you are' right now you know that the 210° bearing/radial is to your right. The easiest thing here would be to turn to a 210° heading to parallel your 'new course.' Before, during, or after your right turn to 210° DON'T CONCERN YOURSELF WITH THE ADF DIRECTION INDICATOR NEEDLE READING. The important thing at the moment is to decide on your intercept angle to your 'new course.' You decide that a 45° angle would be good here. Looking at your Heading Indicator you add 45° to your present heading (210°) and see that your new heading for a 45° intercept of the 210° bearing/radial would be 255°.

Incidentally, there is a 45° mark on most all Heading Indicators so usually no math is involved; just look at the 45° mark and turn to that heading.

After you've made the turn to 255° AND YOU ARE IN WINGS LEVEL FLIGHT watch the ADF Direction Indicator needle move. And when it reaches the mark that is 45° from 18 — the tail of your aircraft — (remember the NDB is over your right shoulder; therefore the ADF Direction Indicator needle will keep moving towards 18 — the tail of your aircraft — as long as you stay on the 255° course — physically look in that direction) make your turn to 210°.

In order not to pass the 210° bearing/radial you must make your left turn to 210° as soon as the ADF Direction Indicator needle reaches the mark that represents 135° on the ADF Direction Indicator dial. That setting (135) is 45° to the right of the tail of your aircraft and is your signal that you should make your left turn to 210°.

When the turn is completed the ADF Direction Indicator needle will point to 18 — the tail of your aircraft — because the NDB is right behind you — as at position #29.

43

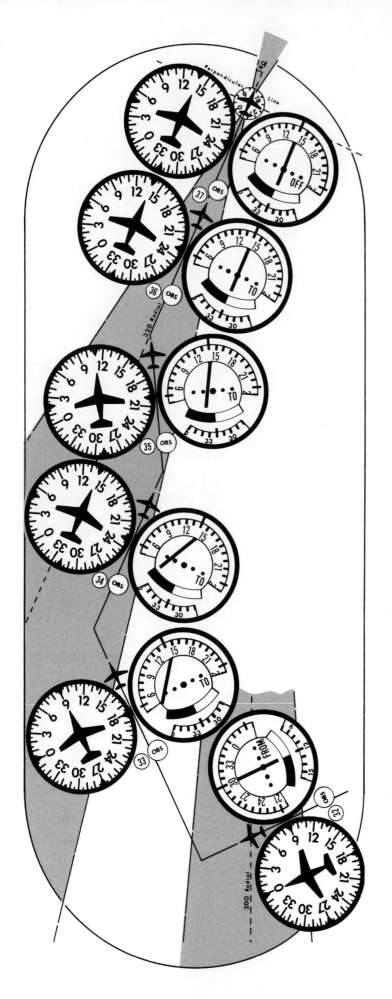

INTERCEPTION OF A RADIAL 'INBOUND' 'TO' THE OMNI INSIDE THE CDI NEEDLE SENSITIVITY AREA

Positions #30 and 31 are not shown here, but they are on the 'in-flight' exercise sheets. They, **and** position #32, are 'radial position' checks and are the only times where I've had you use the FROM reading for a position check.

After your 'radial position' check at #32 you decide to intercept the 320° radial 'inbound' TO the Omni. Now, you want a TO reading so you must dial 14 (140°, Southeast) into your Course Indicator.

If you maintained your present heading you would intercept the 320° radial at an angle greater than 90°. Looking at the Heading Indicator at position #32 you'll see that your 'selected course' is 140° to your right. At position #33 after a 90° right turn you could intercept the 320° radial at a 45° angle and that would be okay. But since these are exercises in different types of interception angles you decide to parallel the 320° radial INSIDE the CDI needle sensitivity area.

Reset your Course Indicator to the 'selected course' 140°, Southeast (14 on your Course Indicator dial). When the CDI needle starts to move toward the center of the dial make a right turn to parallel the 'selected course' and see what intercept your CDI needle suggests. The CDI needle at position #34 suggests an intercept angle of 30°.

The 'selected course' is to your LEFT. Looking at the Heading Indicator at position #34 you'll note that a heading of 110° will be required for the 30° intercept angle.

The CDI needle at position #35 is nearing the center of the dial. Its movement speed is your clue as to when to begin your turn to intercept the 320° radial 'inbound' TO the Omni.

Position #36. You're 'on course' 'inbound' TO the Omni.

Position #37. You're 'on top' the Omni and crossing the perpendicular line. Your Course Indicator reads OFF.

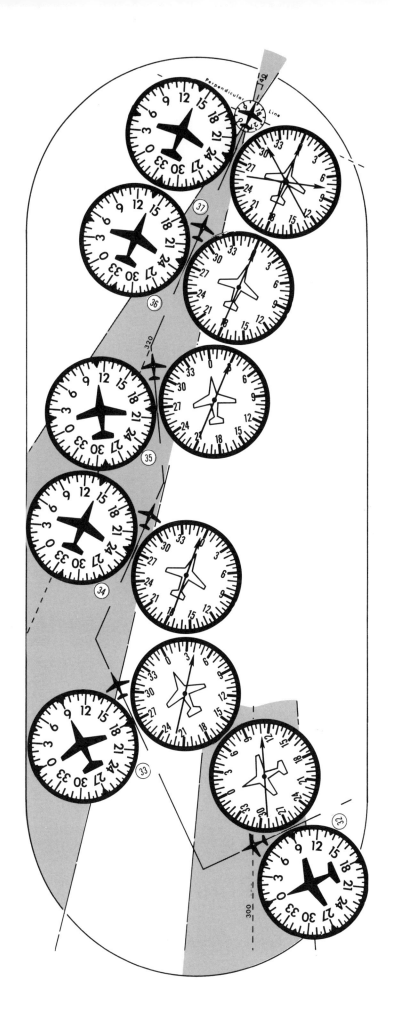

INTERCEPTION OF A BEARING/RADIAL 'INBOUND' TO THE NDB

After tracking 'outbound' on the 210° bearing/radial FROM the NDB you are instructed to make a right turn to a heading of 270°. As soon as you cross the 230° bearing/radial you are to make a right turn to a heading of 360° (0) and maintain that heading until you reach the 300° bearing/radial at which time you make a right turn to 090° and begin the necessary preparations to intercept and track 'inbound' on the 320° bearing/radial TO the NDB.

At position #32 you are aware that you're crossing the 300° bearing/radial because you watched the ADF Direction Indicator needle move until the arrow pointed to 120°. And the tail of the arrow to 300°. This occurred of course because your magnetic direction of flight is the same as the position of your ADF Direction Indicator dial.

At position #34 you are paralleling the 320° bearing/radial. Notice that the ADF Direction Indicator needle is pointing to the area between the nose of your aircraft and your left wing tip (27). The NDB is just left of the nose of your aircraft — make a mental note of that. You're already looking in that direction.

You've decided on a 30° angle of intercept of the 320° bearing/radial. Paralleling the 320° bearing/radial your magnetic heading is naturally 140° (the reciprocal of 320°) so a 30° interception angle would mean a left turn to 110°.

You'll note now that the ADF Direction Indicator needle is now nearing 3 — the mark that is 30° to the right of the nose of your aircraft and equals the 30° interception angle. When it's right on 3 it is time to turn to your new heading of 140°. AFTER YOU'VE TURNED TO 140° AND ESTABLISHED YOUR AIRCRAFT IN WINGS LEVEL FLIGHT look at the ADF Direction Indicator dial and the needle will be pointing to 0 — the nose of your aircraft. The NDB is right in front of you.

At position #37 you are 'on top' the NDB and the needle is starting to swing irratically as you cross the perpendicular line.

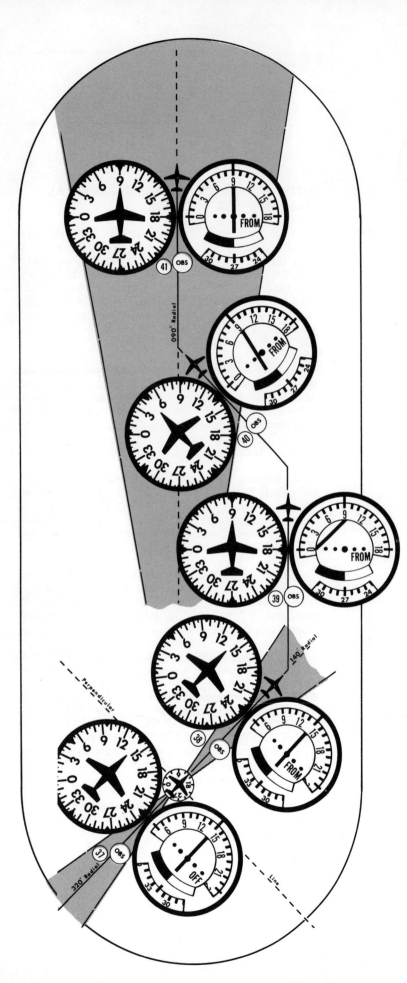

INTERCEPTION OF AN 'OUTBOUND' RADIAL IMMEDIATELY AFTER CROSSING THE OMNI

Position #37 is the same #37 as on page 44. It is shown again for continuity.

Wait until you get a firm FROM reading as at position #38 before turning left to parallel the newly 'selected course' of 090° 'outbound' FROM the Omni.

Resetting the Course Indicator to the 'selected course' of 090° will make the CDI needle 'peg' left because you are outside the CDI needle sensitivity area.

An intercept angle of 45° is good here because it allows you to intercept your 'selected course' quickly.

Again, when the CDI needle starts to move from its 'pegged' position toward center note its speed as this is your clue as to when to start your turn to intercept the 090° radial.

At position #41 you're 'outbound' FROM the Omni.

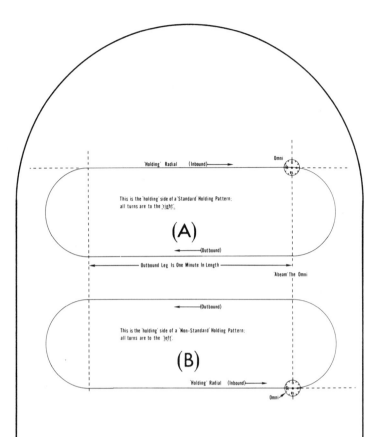

NOTE

The outbound leg of the holding pattern is one minute in length in a **NO-WIND CONDITION** (i.e., the wind is calm). **HOWEVER,** the outbound leg may be more than that sometime due to a headwind so be sure to compensate for this and strive for **A ONE-MINUTE 'INBOUND' LEG OF THE HOLDING PATTERN.**

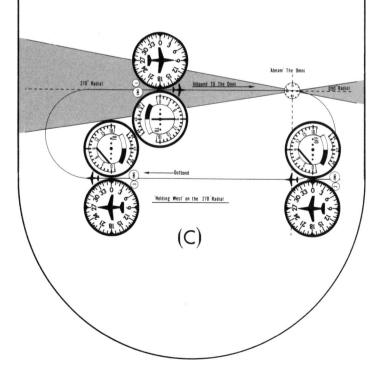

HOLDING

Holding is maneuvering your aircraft in relation to a navigational fix (for our discussion and the 'in-flight' exercises we'll continue to use an Omni) while awaiting further clearance from Air Traffic Control (ATC).

HOLDING PATTERN

The pattern made by your aircraft as you perform the holding maneuver at the Omni. Let's look at the holding pattern from an illustration standpoint first. The two illustrations on your left depict what the 'standard' and 'non-standard' holding patterns look like when flown in a 'no-wind' condition. The 'inbound' and 'outbound' are flown in the same direction respectively in both cases. The holding course (here it's the 270° radial) is the same in both cases. 'Abeam' the Omni is the same in both cases. **THE ONLY DIFFERENCE ARE THE TURNS.** 'Standard' consists of right-hand turns. 'Non-standard' consists of left-hand turns. The 'leg' is one minute long if you're flying below 14,000 feet and one and one-half minutes if above 14,000 feet.

The 'standard' holding pattern is used almost exclusively and only on instructions from ATC (Air Traffic Control) are you to use the 'non-standard' pattern.

Instructions from ATC generally consist of at least these three items; direction **from** the Omni you're to 'hold at,' the 'radial' you're to 'hold on' and whether it's a 'standard' or 'non-standard' holding pattern.

Visualization is very important here due to the fact that the instructions tend to sound a little ambiguous. For instance, the instructions from ATC for the aircraft at C are ' . . . hold West on the 270° radial at the Naperville Omni, one minute right turns, standard . . . ' While you are at position 1 you are 'inbound' TO the Omni. You're holding **West** on the 270° radial. The 'West' they are talking about is the direction **'from'** the Omni. **Visualize** that. In the instructions they call the 'holding course' the radial that is 'inbound' TO the Omni. But your 'course' TO the Omni has to be (in this case) 090° (reciprocal of 270°) in order for you to have a TO reading. The **'COURSE'** you insert into your Course Indicator, based upon the instructions from ATC, is **ALWAYS THE RECIPROCAL OF THE RADIAL THEY TELL YOU TO HOLD ON. REMEMBER THAT!**

When you're at position 2 you're 'abeam the Omni' and going 'outbound.' Notice the OFF indication and the CDI needle reversal (look familiar?). Remember why both have the reading they do? At position 3 the CDI needle is still reversed and one minute has passed since you were 'abeam the Omni' at 2. Now you start your 3° per second turn or 30° bank turn, whichever gives the least angle of bank. In most cases it will be the 3° per second due to your relatively slow speed.

Now let's discuss the best way to intercept the 'radial' on which we are told to 'hold.'

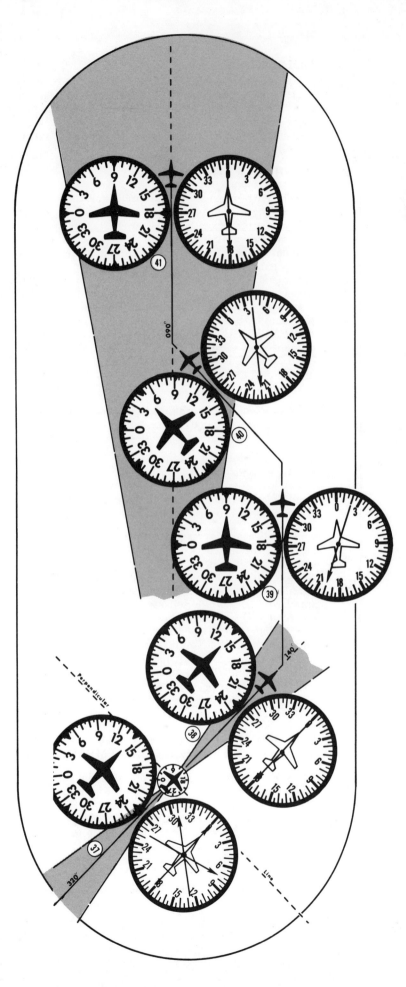

INTERCEPTION OF AN OUTBOUND BEARING/RADIAL IMMEDIATELY AFTER CROSSING THE NDB

Position #38 shows you've crossed the NDB and you are 'outbound' on the 140° bearing/radial and the best way to intercept the 090° bearing/radial and track 'outbound' on it is to first parallel it and decide what angle of intercept you want to use. Here you've decided to use a 45° interception angle. A 45° angle of interception from your present heading of 090° would mean a left turn to 045°. Remember, the NDB is over your left shoulder between your left wing tip and the tail of your aircraft, therefore the 090° bearing/radial, since you're paralleling it, is to your left.

At position #40 you're very close to the interception point and your ADF Direction Indicator needle is almost to the mark that it 45° from the tail of your aircraft. When it reaches that point you are 'on course' and must make your turn to 090°. The NDB is now directly in back of you.

And this ends the illustrations from the Flight Path Sequence Study Sheet. I hope this has helped you understand both the Course Deviation Indicator and the ADF Direction Indicator system of navigation better and how they both rely on the indispensable Heading Indicator.

FLIGHT PATH SEQUENCE STUDY SHEET

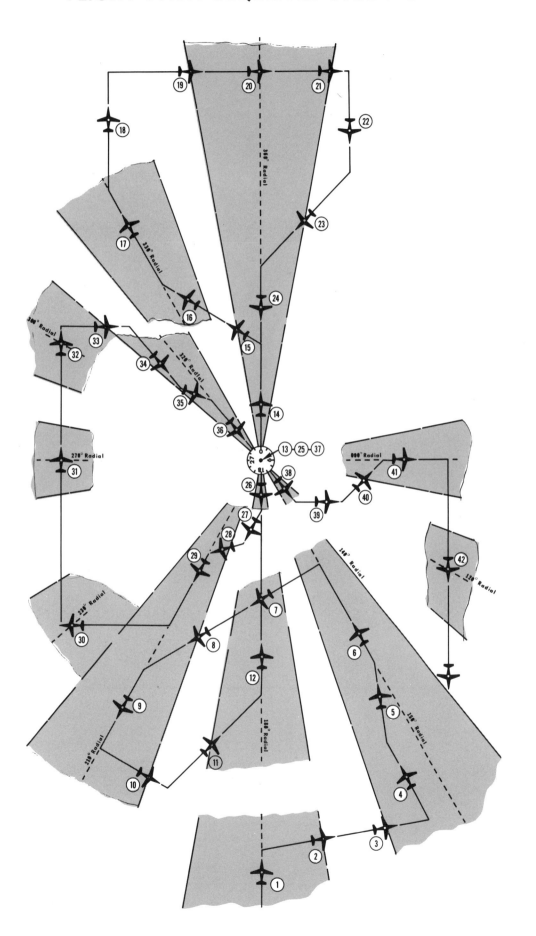

FOLD-OUT

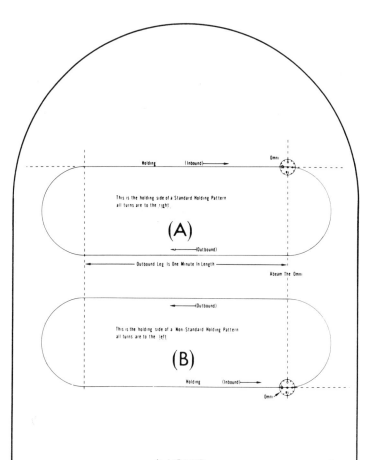

NOTE

The outbound leg of the holding pattern is one minute in length in a **NO-WIND CONDITION** (i.e., the wind is calm). **HOWEVER**, the outbound leg may be more than that sometime due to a headwind so be sure to compensate for this and strive for **A ONE-MINUTE 'INBOUND' LEG OF THE HOLDING PATTERN.**

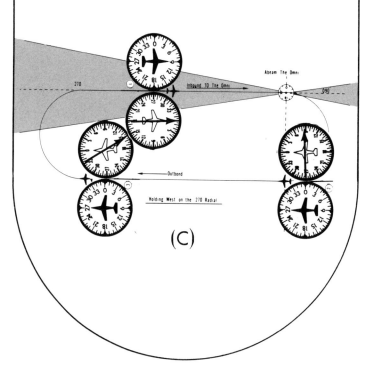

HOLDING

Holding is maneuvering your aircraft in a race-track pattern at an NDB while awaiting further clearance from ATC (Air Traffic Control).

HOLDING PATTERN

The holding pattern made by your aircraft as you perform the holding maneuver at the NDB. Let's look at the holding pattern from an illustration standpoint first. The two illustrations on your left depict what the 'standard' and the 'non-standard' holding patterns look like when flown in a 'no-wind' condition. The 'inbound' and 'outbound' are flown in the same direction respectively in both cases. The holding course (here it is the 270° bearing/radial) is the same in both cases. THE ONLY DIFFERENCES ARE THE TURNS. 'Standard' consists of right-hand turns. 'Non-standard' consists of left-hand turns. The 'leg' is one minute long if you're flying below 14,000 feet and one and one-half minutes if above 14,000 feet.

The 'standard' holding pattern is used almost exclusively and only on instructions from ATC (Air Traffic Control) are you to use the 'non-standard' pattern.

Visualization is very important here too due to the fact that the instructions tend to sound a little ambiguous. For instance, the instructions from ATC for the aircraft at (C) are ' . . . hold west on the 270° bearing/radial at the NDB, one minute right turns, standard . . . ' While you are at position #1 you are 'inbound' TO the NDB. You're **'holding west'** on the 270° bearing/radial. The **'west'** they are talking about is the direction FROM the NDB. **Visualize** that. In the instructions they call the 'holding course' the bearing/radial that is 'inbound TO the NDB.' But your 'course' TO the NDB has to be the reciprocal of 270° — here 090° — in order for you to go TO the NDB.

When you're at position #2 you are 'abeam the NFB' — the ADF Direction Indicator needle is pointing to your right wing tip — and you're going 'outbound.' In contrast to the Course Deviation Indicator of Omni — which points away from the Omni — the ADF Direction Indicator needle continues to point **to** the NDB.

At position #3 you've completed your 'one-minute' 'outbound' and are preparing for your 3° per second turn 'inbound.' DON'T WATCH THE ADF DIRECTION INDICATOR NEEDLE WHILE YOU'RE TURNING. Wait until you've completed your 3° per second turn to 090° and are IN WINGS LEVEL FLIGHT, then make note of it's location.

51

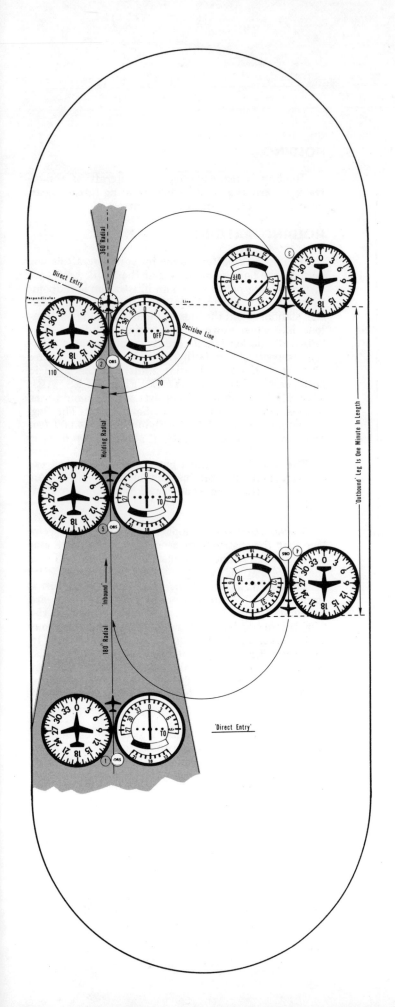

ENTRY INTO THE HOLDING PATTERN

Your heading while you're on your way TO the Omni will determine what type of an entry (direct, parallel or teardrop) you will make into the holding pattern.

The first entry we'll discuss is the . . .

DIRECT ENTRY

The **DIRECT ENTRY** is used whenever your **HEADING TO THE OMNI IS WITHIN 70° TO THE LEFT AND 110° TO THE RIGHT OF THE HOLDING RADIAL.** The instructions you receive from ATC (Air Traffic Control) are ' . . . hold South on the 180° radial, one minute, right turns, standard . . . ' **VISUALIZATION IS EXTREMELY IMPORTANT. MAKE SURE YOU VISUALIZE WHERE THAT 'HOLDING RADIAL' IS IN RELATION TO THE OMNI AT WHICH YOU ARE TO HOLD. AND ALSO VISUALIZE YOUR PRESENT POSITION IN RELATION TO THE OMNI AND THE 'HOLDING BEARING' OMNI.** While at position #1 you note that your 'selected course' TO the Omni is 0 (360° due-North). You have a TO indication so you know already that you are on the 180° radial (due South of the Omni, heading 360°). By looking at your Heading Indicator you see that you are within the **DIRECT ENTRY** decision area, **70° TO THE RIGHT AND 110° TO THE LEFT OF THE 'HOLDING RADIAL.'** In this particular case there is no interception to be made to enter the holding pattern. You're already 'inbound' on the holding radial TO the Omni.

While at position #2 you'll notice that your Course Indicator reads OFF. Why? Because you're 'on top' the Omni and crossing the perpendicular line. As soon as you get a firm FROM reading begin your right, 3° per second, turn to the 'outbound heading of 180° (due South).

When you're abeam the Omni at position #3 on the 180° heading start timing for one minute. When the one minute has expired begin a 3° per second turn to your RIGHT to intercept the 'holding course inbound' TO the Omni.

The Course Indicator setting remains the same as long as you're 'holding' on this particular radial. And because this is so you will encounter the same ambiguities that we discussed in the beginning, namely that of having CDI needle reversal indications. For example, at positions #3 and 4 the CDI needle says that the 'holding course' is to your left. You know from looking at the illustration that the 'holding course' is to your RIGHT. After you begin your RIGHT 3° per second, turn to intercept the 'holding course radial inbound' TO the Omni the CDI needle will begin to move toward the center of your Course Indicator dial **when you reenter the CDI needle sensitivity area.**

The 'on top' the Omni, the reversal of the CDI needle, and the interception of the 'holding radial' is all familiar isn't it? Once you do the 'in-flight' exercises everything will fall into place. You'll find the Holding Patterns to be an extension of what we discussed before.

52

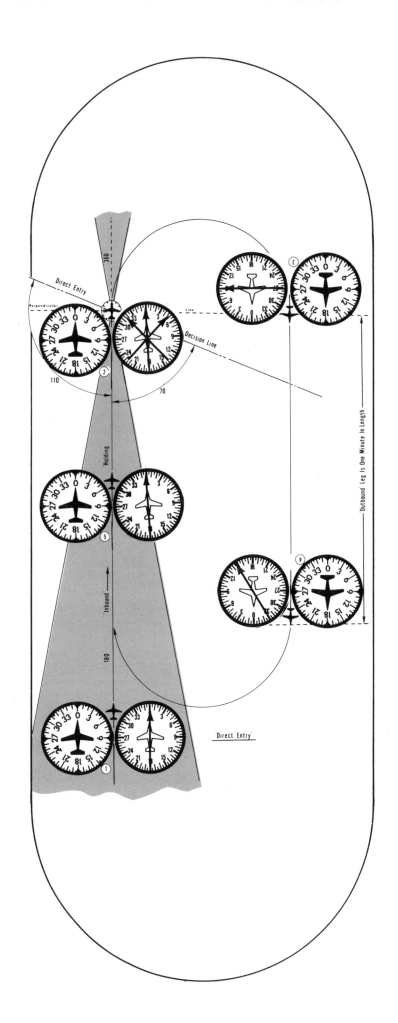

ENTRY INTO THE HOLDING PATTERN

Your heading while on your way TO the NDB will determine what type of entry (direct, parallel or teardrop) you will make into the holding pattern.

The first entry we will discuss is . . .

DIRECT ENTRY

The DIRECT ENTRY is used whenever your HEADING TO THE NDB IS WITHIN 70° TO THE LEFT AND 110° TO THE RIGHT OF THE HOLDING BEARING/RADIAL. The instructions you receive from ATC are ' . . . hold south on the 180° bearing/radial, one minute, right turns, standard . . .' VISUALIZATION IS EXTREMELY IMPORTANT HERE. MAKE SURE YOU VISUALIZE WHERE THAT 'HOLDING BEARING/RADIAL' IS IN RELATION TO THE NDB AT WHICH YOU ARE TO HOLD. AND ALSO VISUALIZE YOUR PRESENT POSITION IN RELATION TO THE NDB AND THE HOLDING BEARING/RADIAL.

While at position #1 you note that your 'selected course' TO the NDB is 0° (360°). The ADF Direction Indicator needle is pointing to 0 — the nose of your aircraft. You're on the 180° bearing/radial going TO the NDB. By looking at your Heading Indicator you see that you are within the DIRECT ENTRY decision area — 70° TO THE RIGHT AND 110° TO THE LEFT OF THE 'HOLDING BEARING/ RADIAL.' In this particular case there is no interception to be made to enter the holding pattern. You're already 'inbound' on the 'holding bearing/ radial' TO the NDB.

While at position #2 you notice your ADF Direction Indicator needle start to swing erratically. That's your signal to be ready to start your right 3° per second turn AFTER THE NEEDLE POINTS TO 18 — THE TAIL OF YOUR AIRCRAFT. When it does, you must concentrate on making the 3° per second turn and rolling out on a heading of 180°. DON'T WATCH THE ADF DIRECTION INDICATOR NEEDLE WHILE TURNING. Establish yourself in WINGS LEVEL FLIGHT and then note where your ADF Direction Indicator needle is pointing. When it points to 9 — your left wing tip, as at position #3 — you are 'abeam the NDB' and it is here when you should start timing for your one-minute 'outbound' leg.

At position #4 you have completed your one-minute 'outbound' leg and will start your 3° per second turn to 360° (0°) 'inbound' TO the NDB. DON'T WATCH THE ADF DIRECTION INDICATOR NEEDLE AS YOU TURN — CONCENTRATE ON YOUR 3° PER SECOND TURN TO 360° (0°). DO NOT LOOK AT IT UNTIL YOU ARE ESTABLISHED IN WINGS LEVEL FLIGHT 'INBOUND' TO THE NDB. At this time your ADF Direction Indicator needle will point to 0 — the nose of your aircraft.

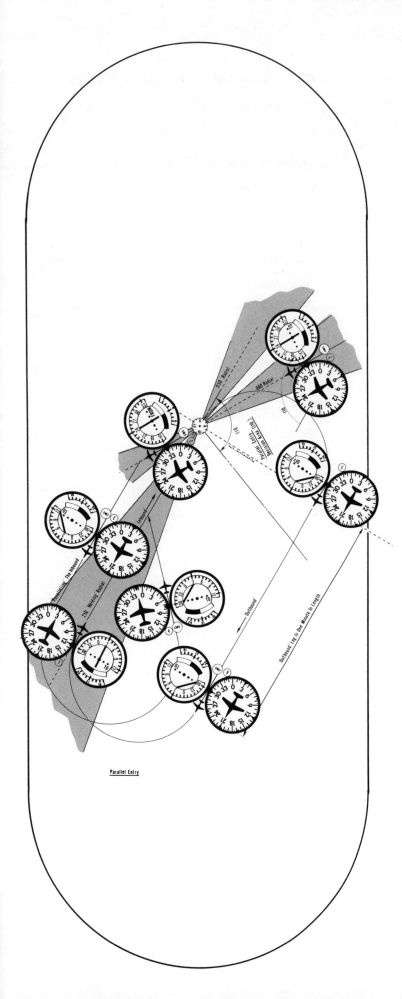

Parallel Entry

PARALLEL ENTRY OMNI

The **PARALLEL ENTRY** is used whenever your **HEADING TO THE OMNI IS WITHIN 110° TO THE LEFT OF THE HOLDING RADIAL.**

The instructions you receive from ATC (Air Traffic Control) in this particular case are ' . . . hold Southeast on the 210° radial, one minute, right turns, standard . . . ' **VISUALIZE WHERE THE 'HOLDING RADIAL' IS IN RELATION TO THE OMNI AT WHICH YOU ARE TO 'HOLD.' THEN VISUALIZE WHERE YOU ARE IN RELATION TO THE OMNI AND THE 'HOLDING RADIAL.'** Visualizing the 'quadrant' the 'holding radial' is in is a good way to do it.

While at position #1 you note that your 'course' TO the Omni is 240° (Southwest) therefore, at the present time you are in the Northeast Quadrant. Are you within the 110° to the RIGHT of the 'holding radial'? Do you see why it is so important to visualize 'where you are' and where you are are to 'hold'? (It will become even more clear as you fly the 'in-flight' exercises.) Looking at your Heading Indicator you note that you ARE within the 110° to the RIGHT of the 'holding radial,' therefore the PARALLEL ENTRY is appropriate.

The PARALLEL ENTRY is so named because the entry into the holding pattern requires that you **PARALLEL THE 'HOLDING RADIAL'** before you intercept it to go 'inbound' TO the Omni.

At position #2 you have a firm FROM reading so you start your LEFT turn to parallel the 'holding radial' for at least one-minute before you make your left, 3° per second, turn 'inbound' TO the Omni.

At position #3 reset your Course Indicator to the 'holding radial course' which is the reciprocal of the 'holding radial.' The 'holding radial' is 210°, but since you want a TO reading you 'select' a course of 030°, Northeast (3° on your Course Indicator dial). The CDI needle is reversed because you are on the TO side of the perpendicular line and, for the moment, you are going away from the Omni. However when you make your turn to intercept the 'holding radial' 'inbound' TO the Omni and you reenter the CDI needle sensitivity area, the CDI needle will move to the left side. It is 'pegged' at position #3 and it will also be 'pegged' at position #4. It will remain 'pegged' until you reenter the CDI needle sensitivity area.

When you're 'on top' the Omni (not shown here because of space) and crossing the perpendicular line the OFF will appear.

The instrument indications at position #5 (abeam the Omni), #6 (end of one minute 'outbound') and #7 ('inbound' TO the Omni) are similar to the direct entry except they are different settings.

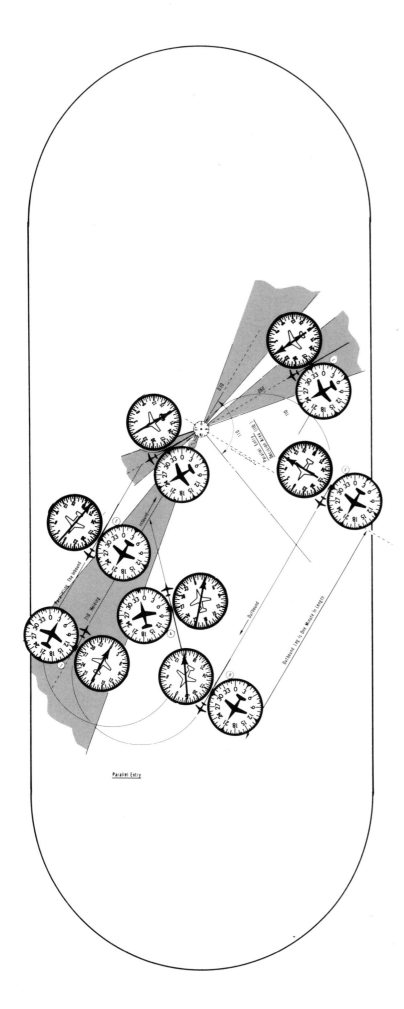

PARALLEL ENTRY NDB

The PARALLEL ENTRY is used whenever your HEADING TO THE NDB IS WITHIN 110° TO THE LEFT OF THE HOLDING BEARING/ RADIAL.

The instructions you receive from ATC in this particular case are '... hold southeast on the 210° bearing/radial, one minute, right turns, standard ...' VISUALIZE WHERE THE 'HOLDING BEARING/ RADIAL' IS IN RELATION TO THE NDB AT WHICH YOU ARE TO 'HOLD.' THEN VISUALIZE WHERE YOU ARE IN RELATION TO THE NDB AND THE 'HOLDING BEARING/RADIAL.' Visualizing the 'quadrant' the 'holding bearing/radial is in, is a good way to do it.

While at position #1 you note that your 'course' TO the NDB is 240° (Southwest). Therefore at the present time you are in the Northeast Quadrant. Are you within the 110° to the RIGHT of the 'holding bearing/radial'? Do you see why it is so important to visualize 'where you are' and 'where you are to hold'? Looking at your Heading Indicator you note that you ARE within the 110° to the RIGHT of the 'holding bearing/radial.' Therefore, the PARALLEL ENTRY is appropriate.

The PARALLEL ENTRY is so named because the entry into the holding pattern requires that you PARALLEL THE HOLDING BEARING/RADIAL before you intercept it to go 'inbound' TO the NDB.

At position #2 your ADF Direction Indicator needle points to 18 — the tail of your aircraft — indicating that you've passed the NDB and you can make your left turn to 210° to parallel that bearing/ radial. Start timing for one minute after you're established on the 210° bearing/radial and at the end of that time make your 3° per second turn to intercept the 210° bearing/radial 'inbound' TO the NDB. It is wise here to make a 45° angle of intercept to the 210° bearing/radial so that you're not too close to the NDB when you roll out onto your 030° heading.

At position #4 you're on an intercept angle of 45° and when your ADF Direction Indicator needle points to the mark that is 45° to the right of your aircraft's nose, that is your clue to turn right to your 'inbound' heading of 030°.

Positions #5, 6 and 7 are the same as the DIRECT ENTRY and TEARDROP ENTRY would be.

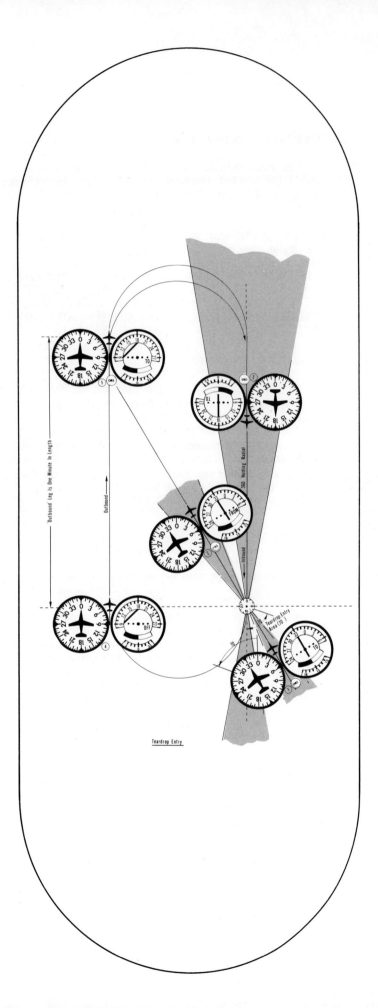

Teardrop Entry

TEARDROP ENTRY (OMNI)

The **TEARDROP ENTRY** is used whenever your HEADING TO THE OMNI IS WITHIN 70° TO THE RIGHT OF THE 'HOLDING RADIAL.'

Your instructions from ATC (Air Traffic Control) for this particular situation are ' . . . hold North on the 360° (0) radial, one minute, right turns, standard . . . ' Again, VISUALIZE WHERE THE 'HOLDING RADIAL' IS IN RELATION TO THE OMNI AT WHICH YOU ARE TO 'HOLD.' AND ALSO VISUALIZE YOUR PRESENT POSITION IN RELATION TO THE OMNI AND THE 'HOLDING RADIAL.' Visualizing the 'quadrant' the 'holding radial' is in and the 'quadrant' you are presently in is a good way to do it.

Your present heading at position #1 is 330° (Northwest). You are in the Southeast quadrant. The 'holding radial' is 360° (due North). By looking at your Heading Indicator you see that you are within 70° to the LEFT of the 'holding radial' so the TEARDROP ENTRY (so named because of its shape) is the most appropriate one.

Have you noticed that **ALL HOLDING PATTERNS ARE THE SAME? THE ONLY DIFFERENCE IS THE ENTRY THAT YOU MAKE INTO THEM.**

Always make sure you have a firm FROM reading before resetting your Course Indicator to the 'inbound holding course.'

After you cross over the 'top of the Omni' at the beginning of your entry into the 'holding pattern' maintain the same heading for at least one minute before you start your right, 3° per second, turn to intercept the 'holding radial inbound' TO the Omni. Here again you will have the CDI needle reversal when you reset your Course Indicator after you cross the Omni so understand why it is happening and **DON'T PANIC.** The CDI needle indication is right based on the data you've inserted into the Course Indicator.

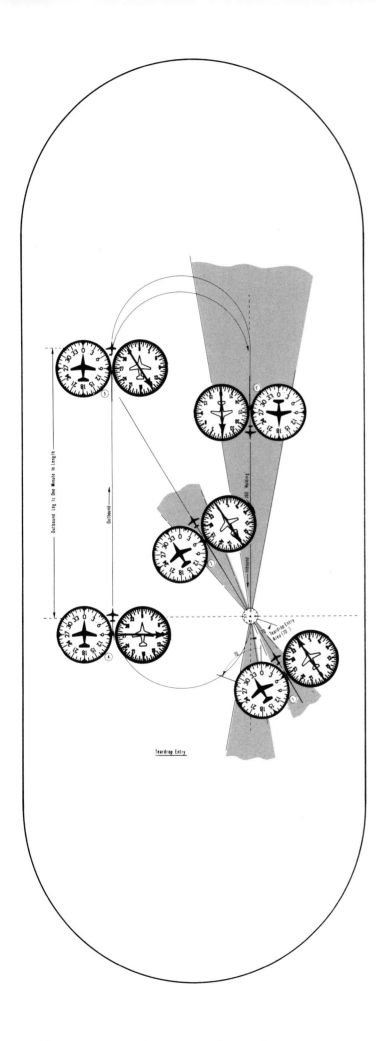

TEARDROP ENTRY (NDB)

The TEARDROP ENTRY is used when your HEADING TO THE NDB is within 70° TO THE RIGHT OF THE 'HOLDING BEARING/RADIAL.'

Your instructions from ATC for this particular situation is '... hold north on the 360° bearing/radial, one minute, righ turns, standard ...' Again, VISUALIZE WHERE THE 'HOLDING BEARING/RADIAL IS IN RELATION TO THE NDB AT WHICH YOU ARE TO 'HOLD' AND ALSO VISUALIZE YOUR PRESENT POSITION IN RELATION TO THE NDB AND THE 'HOLDING BEARING/RADIAL.' Visualizing the quadrant the 'holding bearing/radial' is in and the 'quadrant' you are presently in, is a good way to do it.

Your heading at position #1 is 330° and you are in the southeast quadrant going TO the NDB.

At position #2 you've crossed the NDB still on the same heading — 330°. When the ADF Direction Indicator needle pointed to 18 — the tail of your aircraft — you started timing for one minute and then began your 3° per second right turn to a heading of 180°. DON'T WATCH THE ADF DIRECTION INDICATOR WHILE YOU ARE TURNING. Check it after you've ESTABLISHED YOURSELF IN WINGS LEVEL FLIGHT on the 180° heading.

At position #3 you're 'inbound' TO the NDB.

At position #4 you're 'abeam the NDB and at #5 you're at the end of the one-minute 'outbound leg' and are ready to make the 3° per second turn 'inbound' TO the NDB.

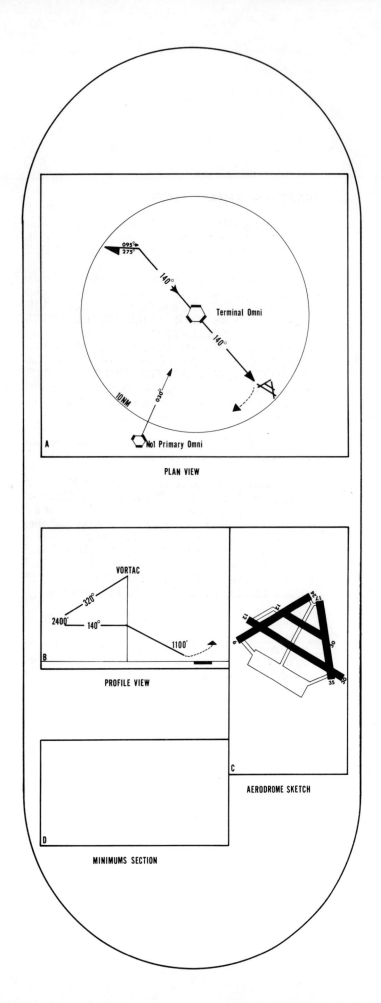

PLAN VIEW

PROFILE VIEW

AERODROME SKETCH

MINIMUMS SECTION

OMNI APPROACH

An Omni approach to an airport consists of:
1. interception of the required radial
   a. to perform the procedure turn or
   b. by following a radar vector or
   c. performing a holding pattern in lieu of a procedure turn.
2. the final approach (which will be either a straight-in or circling approach)
3. visual sighting of the approach lights, runway lights or the runway at, or before, the time you reach the minimum descent altitude (MDA) and be in a position to land.
4. a missed approach if the approach lights, runway lights or the runway are not visible when you reach the minimum descent altitude (MDA)

The approach chart illustration to your left is void of all the criteria found on an actual approach chart because I want to discuss with you the similiarities that exist between the Omni and the ILS approaches plus show how both are related in regards to their orientation and interception. (For a complete detailed explanation of the approach chart I suggest you contact Jeppesen or the National Ocean Survey (NOS) charts.

The top portion (A) of the approach chart of both the Omni (to your left) and the ILS (page 67) is a plan view of the type of approach that is to be performed.

The Omni approach chart gives you the heading from the primary approach Omni (here it is 030°) TO the terminal Omni (so named because it is at the termination of your flight), the headings of the procedure turn (here they are 275 and 095°) and the headings 'inbound' To the Omni (here it is 140°) and the heading to maintain (also 140°) until you reach the minimum descent altitude (MDA). It also shows the Omni to which you must return (here #1) in case you cannot complete a successful landing, i.e., execute a missed approach. The top portion of the ILS chart displays the same thing with these exceptions, a. the heading from the initial approach point to the **outer marker** (LOM), b. includes the 'outbound' heading **from** the outer marker (LOM). Everything else is pretty much the same.

The B portion of the approach chart is a profile view of the A portion. Here you are given the 'outbound' **and** the 'inbound' heading. It also tells you whether the Omni is a VOR or a VORTAC and gives you the altitude for the procedure turn (here it is 2400') and the missed approach altitude (here it is 1100'). The ILS does the same except the middle marker (MM) is the missed approach point (MAP) whereas 1100' on the Omni chart is the minimum descent altitude (MDA) which is also the missed approach point (MAP). Both serve the same purpose.

The C portion of both are the same in that they depict the airport plain view in detail.

The D portion of both give the altitude and distance minimums for that particular airport for both straight-in and circling approaches.

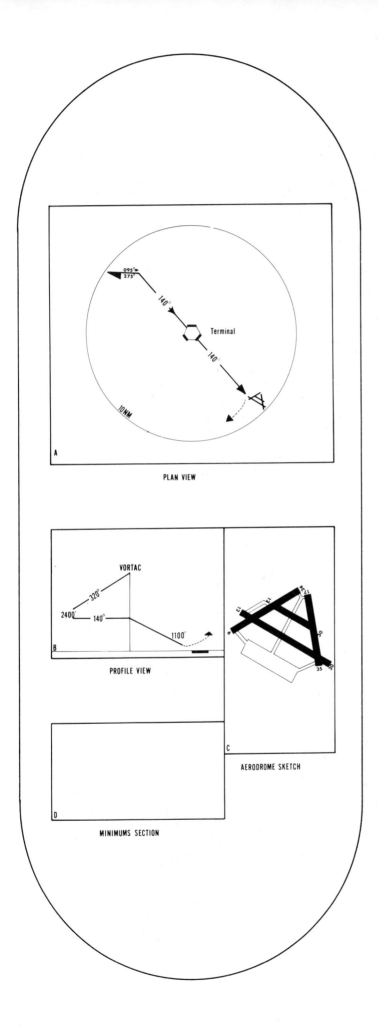

PLAN VIEW

PROFILE VIEW

AERODROME SKETCH

MINIMUMS SECTION

NDB APPROACH

An NDB approach to an airport consists of:
1. interception of the required bearing/radial
   a. to perform the procedure turn or
   b. by following a radar vector or
   c. performing a holding pattern in-lieu-of a procedure turn
2. the final approach (which will be either a straight-in or a circling approach)
3. visual sighting of the approach lights, runway lights or the runway at, or before, the time you reach the minimum descent altitude (MDA) and be in a position to land.
4. a missed approach if the approach lights, runway lights or the runway are not visible when you reach the minimum descent altitude (MDA).

The approach chart illustration to your left is void of all the criteria found on an actual approach chart because I want to discuss with you the similarities that exist between the NDB and the ILS approaches plus show how both are related in regards to their orientation and interception. (For a complete detailed explanation of the approach chart I suggest you contact Jeppesen or the National Oceanic Survey (NOS) charts).

The top portion (A) of the approach chart of both the NDB and the ILS (page 67) is a plan view of the type of approach that is to be performed.

The NDB approach chart gives you the heading from the primary approach NDB TO to the terminal NDB. The headings of the procedure turn (here they are 275° and 095°) and the headings 'inbound' TO the NDB (here it is 140°) until you reach the minimum descent altitude (MDA). It also shows the NDB to which you return in case you cannot complete a successful landing. The top portion of the ILS chart displays the same thing with these exceptions, a. the heading from the initial approach point to the **outer marker** (LOM), and b. includes the 'outbound' heading FROM the outer marker (LOM). Everything else is pretty much the same.

The B portion of the approach chart is a profile view of the A portion. Here you are given the 'outbound' **and** 'inbound' heading. It also gives you the altitude for the procedure turn (here it is 2400') and the missed approach altitude (here it is 1100'). The ILS does the same except the middle marker (MM) is the missed approach point (MAP) whereas 1100' on the NDB chart is the minimum descent altitude (MDA) which is also the missed approach point (MAP). Both serve the same purpose.

The C portion of both are the same in that they depict the airport plan view in detail.

The D portion of both give the altitude and distance minimums for that particular airport for both straight-in and circling approaches.

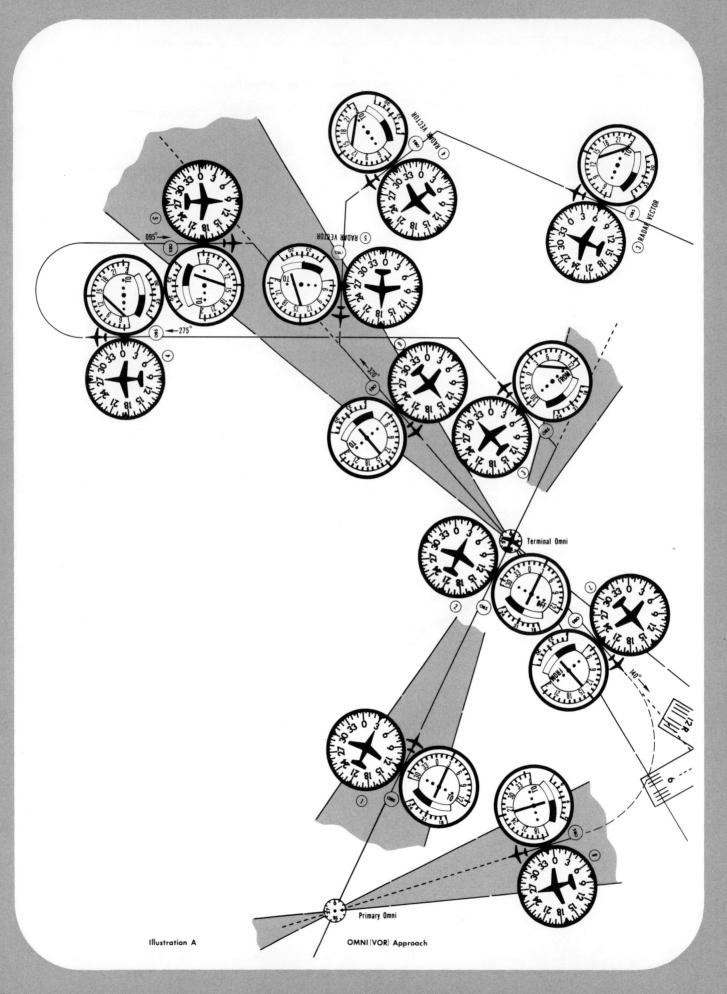

Illustration A — OMNI (VOR) Approach

PROCEDURE TURN (OMNI)

A procedure turn is a maneuver designed to align your aircraft on an 'inbound' course TO the Omni in the final approach configuration in preparation for a landing at your destination airport.

It is a required maneuver except 1), when the symbol NoPT (no procedure turn) is shown, 2), when radar vectoring is provided 3), when a one minute holding pattern is published 'in-lieu-of' procedure turn or 4) when a procedure turn is not authorized.

It is depicted on the plan view of the approach chart as a barb (➤). The example on the foregoing page shows the barb, but the example on your left shows the actual flight path of a procedure turn (assuming a no-wind condition).

You know what 'entry into the holding pattern' is so we'll discuss the 'entry into the procedure turn' first. Then we'll see how that leads into the final approach, the straight-in or circling approach **and** the missed approach, in case that becomes necessary for you to perform at anytime.

Your entry into the procedure turn is governed by your heading upon arrival at the Omni. THE MOST IMPORTANT THING TO REMEMBER IS TO ALWAYS TURN IN THE SHORTER DIRECTION (this could be either right or left of your present direction) IN ORDER TO INTERCEPT THE 'OUTBOUND' COURSE **FROM** THE OMNI WITH AT LEAST A 30° INTERCEPT ANGLE. You'll notice on the illustration at left that I DID NOT show the intercept of the 'OUTBOUND' RADIAL FROM the Omni. You know by now how to intercept radials so I LEFT IT OUT FOR CLARITY. The 'outbound' leg of the procedure turn is usually 2 to 3 minutes in length so you have time to intercept the 'outbound' leg and get yourself 'set-up' for the procedure turn no matter what direction your flying when you cross the Omni.

The procedure turn must be performed within a prescribed area which is usually a ten mile radius from THE CENTER OF THE OMNI (that's the circle you see drawn around the Omni, and the outer marker (LOM) on the ILS approach chart).

The approach chart as you know also depicts a profile view of the procedure turn with the instructions as to when to begin your descent from the procedure turn altitude on your final approach to the airport runway. The text and the illustrations will discuss and show the descent-from-altitude both 'before the Omni' and 'at the Omni.' However, in the pictorial part of the 'in-flight' exercises I have dispensed with the 'descent-from-altitude' part of the procedure turn and final approach because I think it is better that you understand how to intercept the given headings, perform the procedure turn, the final approach and the missed approach in compass direction only, then later add the 'descent-from-altitude' to the 'in-flight' exercises.

Illustration A, to your left, shows your entry into the procedure turn to be from the primary Omni in the Southwest Quadrant.

At #1 you're going from the primary Omni TO the terminal Omni. As before you're on the 210° radial on a 'selected course' of 030° (3 on the Course Indicator dial). You're on the TO side of the perpendicular line represented here by the 300 and 120° radials and you're going TO the Omni.

At #2 you're 'on top' the Omni (sound familiar?). At #3 you're paralleling the 'outbound' heading after having gotten a firm FROM reading on the Course Indicator. Also at #3 the Course Indicator is still set to 3 (030°). After changing to the new heading of 320° you reset the Course Indicator to 32 (320°) in order to intercept the 'outbound' heading for your required 2 to 3 minute outbound flight (that intercept is not shown because it would have made the illustration too overloaded). After intercepting the 320° radial 'outbound' for 2 to 3 minutes you make your left turn to 275° (#4) and fly that heading for approximately one minute (assuming no-wind) and at the same time resetting your Course Indicator to 14 (140° your 'inbound' heading). At #5 you've made your 180° turn to 095° in preparation for interception of the 320° radial 'inbound' TO the Omni. The CDI needle is where it should be because your 'selected course' is to your left. At #6 your 'inbound' TO the Omni and on the TO side of the perpendicular line represented here by the 070 and 240° radials. At #7 you're on the FROM side of the perpendicular line having just crossed the Omni 'outbound.' Where #7 is is usually the point (missed approach point-MAP) at which you must either land or if that's impractical you must execute a missed approach and at #8 fly back TO the primary Omni and await instructions from ATC (Air Traffic Control).

The illustration in the upper right-hand corner (3RV, 4RV, and 5RV) are examples of what might transpire if you were 'radar vectored' to the 320° radial. The radar controller will usually put you in a position so that you can intercept the required radial with an intercept angle between 30 and 45° (here the intercept angle is 35°).

Have your Course Indicator properly set so that you know 'where you are' in relation to the radial and you'll be prepared when the radar controller 'let's you go.' REMEMBER, THE RADAR CONTROLLER VECTORS YOU TO A POINT WHERE YOU CAN INTERCEPT THE 'INBOUND' RADIAL ON YOUR OWN, SO VISUALIZE 'WHERE YOU ARE' AT ALL TIMES.

Note that the runway is slightly to your left (18° in this case). ALWAYS BE PREPARED FOR THIS TYPE OF SITUATION. A STRAIGHT-IN APPROACH DOES NOT NECESSARILY MEAN THAT THE RUNWAY IS ALWAYS STRAIGHT-IN-FRONT-OF-YOU.

Illustration A

OMNI (VOR) Approach

PROCEDURE TURN USING THE NDB

A procedure turn using the ADF Direction Indicator system of navigation is performed the same way as when using the Course Deviation Indicator system of navigation. Instead of using all NDBs in the illustration I've left the Primary Omni there and changed the Terminal Omni to a NDB because that may be the type of situation you will encounter sometime.

At position # 1 you've crossed the Primary Omni using the Course Deviation Indicator system of navigation and your 'course' FROM the Omni TO the NDB is 030° so the Heading Indicator reading doesn't change and your ADF Direction Indicator needle point to 0 — the nose of your aircraft — at position #1A.

At position #2 you're 'on top' the NDB and the ADF Direction Indicator needle has begun to swing erratically indicating imminent NDB passage.

After crossing the NDB and the needle has swung to the tail position of your aircraft — 18 — you can make the left turn to 320° to parallel the 'inbound' course of 140° TO the NDB at position #3. DON'T WATCH THE ADF DIRECTION INDICATOR NEEDLE ONCE YOU'VE STARTED THE TURN TO THE HEADING OF 320°. Address yourself to keeping the magnetic heading of 320° for 2 to 3 minutes and while you're doing that not that your next turn will be a heading of 275° and also note that 275° from 320° is 45°. THAT'S IMPORTANT BECAUSE YOU'LL WANT TO KNOW WHEN YOU CROSS THE 320° BEARING/RADIAL. Since your direction of flight will be 275° and you're crossing the 320° bearing/radial at the 45° angle the point at which the ADF Direction Indicator needle points to the mark on the dial that is 45° from 275° that's your clue to having crossed the 320° bearing/radial, i.e., the mark on the dial that represents 45° from 275° is 230°. But that 230° DOES NOT MEAN A THING AS FAR AS MAGNETIC DEGREES ARE CONCERNED, THAT'S JUST THE POINT ON THE ADF DIRECTION INDICATOR DIAL THAT SHOWS YOU'VE CROSSED THE 320° BEARING/RADIAL. (In this instance the aircraft is depicted **behind** the dials due to space limitations.)

After crossing the 320° bearing/radial you fly another minute (assuming a no-wind condition) before making your right 3° per second 180° turn to the 095° heading. (Once you've crossed the 320° bearing/radial it is best not to look at the ADF Direction Indicator dial until you're in WINGS-LEVEL FLIGHT on the 095° heading.) Your heading to the 320° bearing/radial is 095° and since that parallels your 275° heading you're going to intercept the 320° bearing/radial as a 45° angle. Here again you must use your Heading Indicator — which you properly set according to your Magnetic Compass while you were cruising in WINGS-LEVEL FLIGHT while paralleling the 320° — because without is you'll never know when to make your turn to intercept the 320° bearing/radial at a 45° angle. The needle of the ADF Direction Indicator is pointing to an area which is to the right of your aircraft's nose and ahead of your right wing tip — physically point in that direction to orient yourself. You know you are going to be turning right and you know that you are intercepting the 320° bearing/radial at 45°. And after you make your turn to 140° (the reciprocal of 320°) you will want the ADF Direction Indicator needle to point at 0 — the nose of your aircraft — so when the ADF Direction Indicator needle points to the mark that is 45° to the right of your aircraft's nose — 0 on the dail — you know you are on the 320° bearing/radial and it is time to turn 'inbound.' Adding 095° to 45° will give you 140° — your heading TO the (RBn) (NDB).

DON'T WATCH THE ADF DIRECTION INDICATOR NEEDLE WHILE TURNING, check it after you are in WINGS-LEVEL FLIGHT to see if you have to make any course corrections.

At position #7 the needle is pointing to 18 — the tail of your aircraft — and if you can't see the runway and have to make a missed approach your Course Deviation Indicator is all set up and you return to the Omni for further instructions.

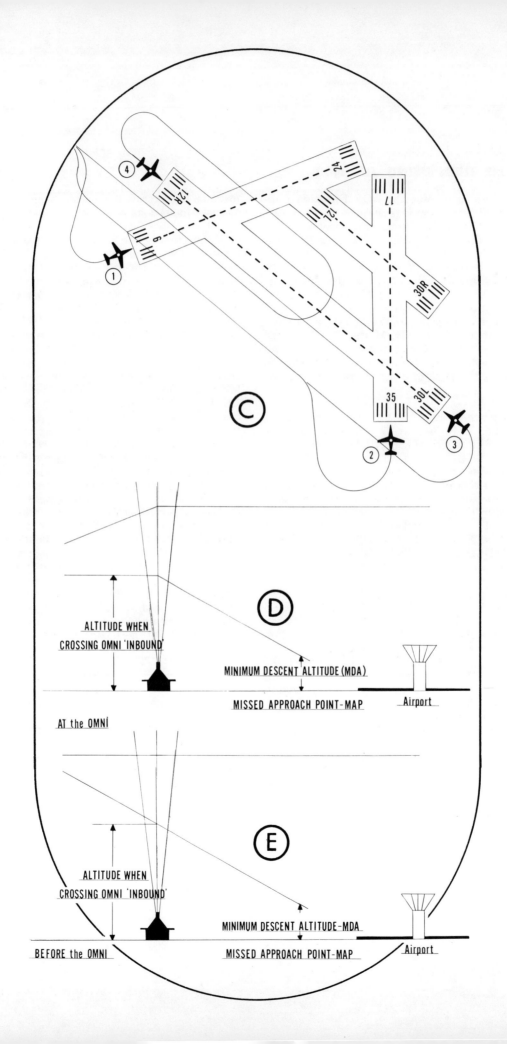

APPROACHES: STRAIGHT-IN, CIRCLING, MISSED AND MISSED APPROACH POINT

A STRAIGHT-IN APPROACH DOES NOT NECESSARILY MEAN THAT THE RUNWAY IS ALWAYS IN FRONT OF YOU. IT MAY BE AS MUCH AS 30° TO EITHER SIDE OF YOUR APPROACH COURSE. BE PREPARED FOR THAT WHEN YOU DO AN ACTUAL OMNI OR NDB APPROACH. See illustration B.

Illustration C depicts an airport and the different ways you could perform the CIRCLING APPROACH. The CIRCLING APPROACH is a visual flight manuever and is used to help align your aircraft with the landing runway. Each landing situation is different because of the variables of ceiling, visibility, wind direction and velocity, obstructions and final approach course. Since these variables exist in many conditions there is no set procedure for accomplishing the CIRCLING APPROACH under every situation. It is possible that you may have to perform a right-hand traffic pattern but try, when possible, to always do a left-hand traffic pattern so you can keep the runway in sight at all times Be cautious about making steep turns when close to the ground. CIRCLING APPROACHES ARE NOT RECOMMENDED FOR AIRCRAFT THAT CANNOT SAFELY MANUEVER WITHIN THE AREA THAT PROVIDES FOR SAFE OBSTRUCTION CLEARANCE.

Numbers 1, 2 and 3 could be used as circling approaches to runways 6, 35 and 30L. Number 4 could be used if you find yourself too high on the straight-in approach or if you sight the runway too late.

IF AT ANY TIME, WHETHER FROM A STRAIGHT-IN OR CIRCLING APPROACH, YOU LOSE SIGHT OF THE RUNWAY YOU MUST EXECUTE A MISSED APPROACH, EVEN THOUGH ON THE CIRCLING APPROACH YOU MAY HAVE PASSED THE MISSED APPROACH POINT (MAP).

A MISSED APPROACH is, as the name suggests, an approach you couldn't complete because you couldn't visually 'see' the approach lights, runway lights or the runway. The approach chart or plate tells you where to go when executing a missed approach. Remember that when you are doing the Omni or the NDB approach. After you have yourself **and** your aircraft under control contact Air Traffic Control (ATC) and tell them that you executed the missed approach and you are requesting further clearance. BUT MAKE SURE YOU HAVE YOUR AIRCRAFT UNDER CONTROL BEFORE YOU CONTACT ATC. Once you have contacted them they will tell you what to do.

Intercepting bearings/radials while performing Omni and NDB approaches is the same as intercepting bearings/radials while flying TO or FROM and Omni or NDB anytime you are flying in an aircraft or a flight simulator. One thing has been added to the Omni and NDB approaches is that you are required, at a certain specified place (noted on the approach chart profile view) when on final approach to descend from your present altitude to the MISSED APPROACH POINT (MAP) which is the MINIMUM DESCENT ALTITUDE (MDA).

Of the two illustrations at left, D depicts the 'descent-from-altitude' to be AT THE OMNI OR THE NDB. That means that when you're 'on top' the Omni or the NDB you start descending toward the airport at which you intend to land. One of two things will happen: either you'll be able to visually 'see' the approach lights, runways lights or the runway at the MISSED APPROACH POINT (MAP) which on an OMNI or NDB APPROACH is the MINIMUM DESCENT ALTITUDE (MDA) and therefore make a successful landing **or** you'll NOT be able to visually 'see' the runway when you reach the MINIMUM DESCENT ALTITUDE (MDA) and you'll have to execute a MISSED APPROACH.

NOTE: You should descend from the final approach fix to the MDA at 700 to 800 feet per minute.

Illustration E shows the descent beginning when you finish your procedure turn and, depending on the airport at which you're landing, will specify a particular height you must be at when you cross the Omni or the NDB 'inbound on your final approach.' Illustration E, like D, shows MINIMUM DESCENT ALTITUDE (MDA) and the MISSED APPROACH POINT (MAP) to be the same. As before, you land if you can visually 'see' the runway when you reach the MINIMUM DESCENT ALTITUDE (MDA) and you execute a MISSED APPROACH when you CANNOT VISUALLY 'SEE' THE RUNWAY.

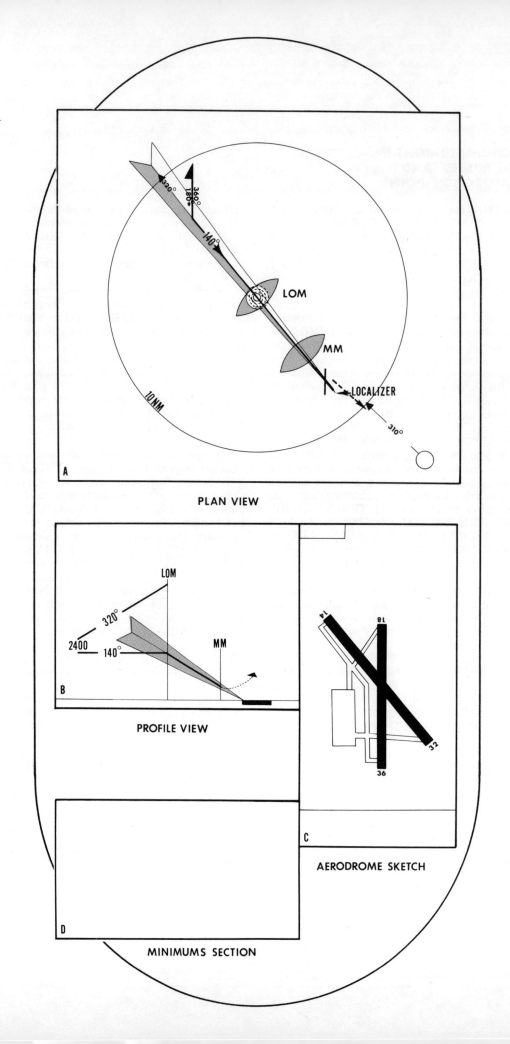

A PLAN VIEW

B PROFILE VIEW

C AERODROME SKETCH

D MINIMUMS SECTION

ILS APPROACH

An ILS (Instrument Landing System) approach to an airport consists of:
1. interception of the required course for the procedure turn or a radar vector for an intercept of the 'inbound' course (localizer).
2. the final approach, which will be a straight-in or circling approach.
3. visual sighting of the approach lights, runway lights or the runway at, or before, the time you reach the missed approach point (MAP); the point at which you execute the missed approach. The decision height (DH) or the minimum descent altitude (MDA).

That sounds like the Omni or the NDB approach doesn't it? There are some differences. The main difference is that an ILS (Instrument Landing System) approach is much more precise. The ILS is designed to provide you with an approach path for EXACT ALIGNMENT AND DESCENT on final approach to your landing runway. Precise control of your aircraft, of course, depends on your correct interpretation of the information given you on your aircraft instruments (namely your Course Deviation Indicator, Heading Indicator, Glide Slope and Altimeter). ILS, like the Omni and the NDB allows you to land under low ceilings and low visibility conditions but with a more precise flight path than an Omni or NDB approach. The ILS makes it more precise in providing you with guidance information (right or left of the centerline or the runway) glide slope information (too high or too low of an angle of descent to your landing runway) and range information (how close you are to your touchdown point and/or your missed approach point — MAP).

The approach chart of the Omni and the NDB and the ILS look very similar. The main difference in the plan view is the final approach. The Omni and the NDB approaches show the Omni and the NDB as the midpoint and the ILS approach shows the outer marker (LOM) as the midpoint of the 10 mile radius in which the approach manuever is to be performed. Sometimes the NDB is located on the airport.

The profile views in both are similar too, except the ILS shows the slide slope in profile whereas the Omni and NDB show just the flight path alone.

To be thoroughly familiar with all the information contained on the approach plates either contact Jeppesen or the NOS (National Oceanic Survey). Both have pamphlets explaining in detail the various information presented on the approach charts.

The ILS approach chart, as does the Omni and the NDB approach charts, provides you with the necessary references for your 'outbound' heading from the outer marker (LOM), your correct headings for the procedure turn and your correct headings for the final and where to go from the missed approach point (MAP), in case that becomes necessary for you to perform.

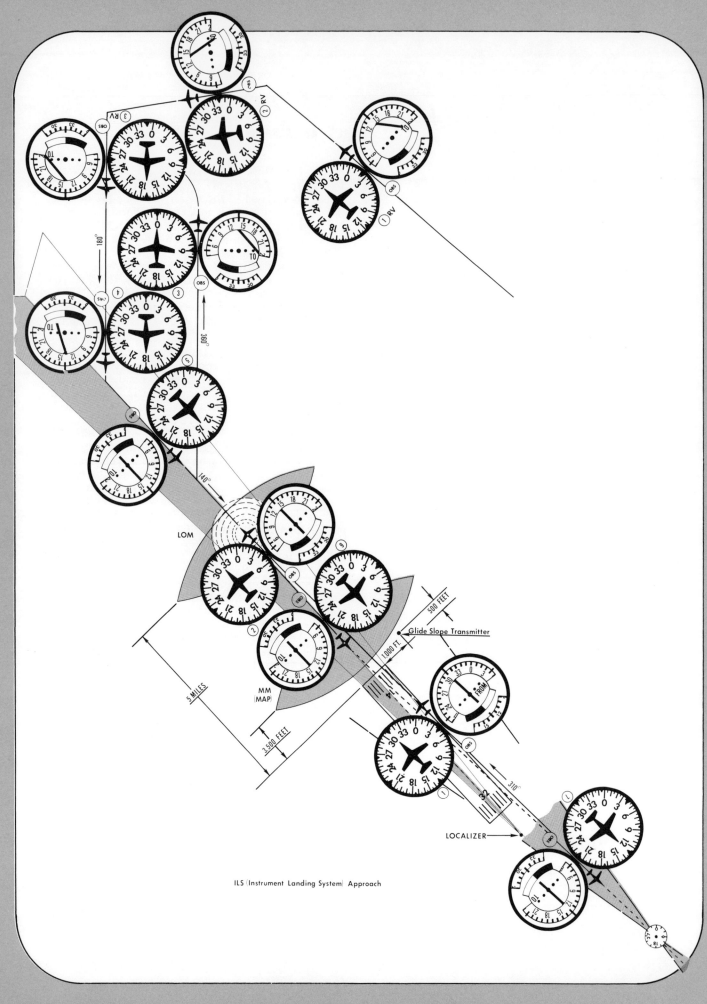

ILS (Instrument Landing System) Approach

PROCEDURE TURN

We'll discuss the procedure turn first as we did with the Omni approach to show the similarity between the two. Then how that leads into the final approach (straight-in or circling) **and** the missed approach.

The procedure turn is the maneuver prescribed when it is necessary to reverse direction to establish your aircraft 'inbound' on a final approach course. It is a required maneuver except when the symbol NoPT (no procedure turn) is shown, when radar vectoring is provided; when a one minute holding pattern is published 'in lieu of' a procedure turn, or when a procedure turn is not authorized.

As with the procedure turn during an Omni approach the ILS procedure turn maneuver must be completed in the prescribed area and on the headings noted on the approach chart.

The illustration to your left shows your flight path to the outer marker (LOM) is on the 310° radial FROM the Omni. Since the outer marker (LOM) does not radiate radials like the Omni, you must make sure you maintain an exact course, regardless of wind, FROM the Omni to the outer marker (LOM). Once the outer marker light lights up on your instrument panel you begin your turn to the 'outbound' heading and maintain that for one (1) minute.

You've probably noticed by now the difference in the Course Indicator being presented here. This is the only type you can use for an ILS approach. You must have one needle (the vertical one) to show whether you're right or left of your course and the other needle (the horizontal one) to show whether you're too high (above the glide slope flight path) or too low (below the glide slope flight path).

One reason that an ILS approach is more precise is the fact that the vertical and horizontal needles are more sensitive due to the frequency modulation that is built into the ILS system. This means that even the slightest deviation from either the horizontal or the vertical flight path will be noticed immediately. For instance the horizontal dots on your Course Indicator when performing an ILS approach are representative of 1¼ degrees compared to 5 and 10° when using the Course Indicator for an Omni approach. And the vertical dots each represent ½ degrees. As an example; if the vertical needle of your Course Indicator shows a one dot needle deflection when you are one mile from the touchdown point of your landing runway you are approximately 300 feet right or left (whichever side the needle is on) of the center-line of the runway. Enough to put you in the grass beside the runway unless you correct the error. Likewise a one dot horizontal needle defelection on your Course Indicator below the large dot in the center of the dial means that at one mile from your touchdown point you are approximately 25 feet lower than what you want to be, and that's not good.

The ground equipment required to give you this precision on an ILS approach consists of 1), the outer marker (LOM) and the middle marker (MM), which are called marker beacons, 2), a localizer transmitter located on the opposite end of the approach end of the ILS runway which helps you keep aligned with the centerline of the runway and 3), the glide slope transmitter located 1,000 feet to the side and 500 feet from the end of the runway which helps you keep on the proper glide slope (angle of descent) to your landing runway.

At position #1 you're 'outbound' FROM the Omni on the 310° radial. At #2 you're 'on top' the outer marker (LOM) and setting yourself up for the one minute flight 'out-bound' from the outermarker (LOM). At #3 you're on the first turn of the procedure turn and after approximately one minute of flight (assuming no wind) you make a 180° left turn to intercept the ILS localizer path 'inbound' (#4). Notice how far to the right the CDI needle is at #4, yet you are ready to make your turn to intercept the localizer path 'inbound.' The CDI NEEDLE IS MUCH MORE SENSITIVE WHEN USING THE ILS than it is when using the Omni. It is 4 times more sensitive than Omni.

At #5 you're 'inbound' and almost to the outer marker (LOM), your first clue as to how far away you are from the runway. At #6 you're at the middle marker (MM) and 3,500 feet away from the runway. This (#6) is also your missed approach point (MAP) and this is where you execute a missed approach if you cannot see the approach lights, runway lights or the runway.

1RV, 2RV and 3RV are examples of what might transpire if you were radar vectored to the localizer course. Interception of the localizer course from a radar vector 'hand-off' is just like the Omni radar vector EXCEPT THE ILS IS MUCH MORE SENSITIVE so you really have to be 'on your toes' when intercepting the ILS from a radar vector.

Position #7 simulates a missed approach.

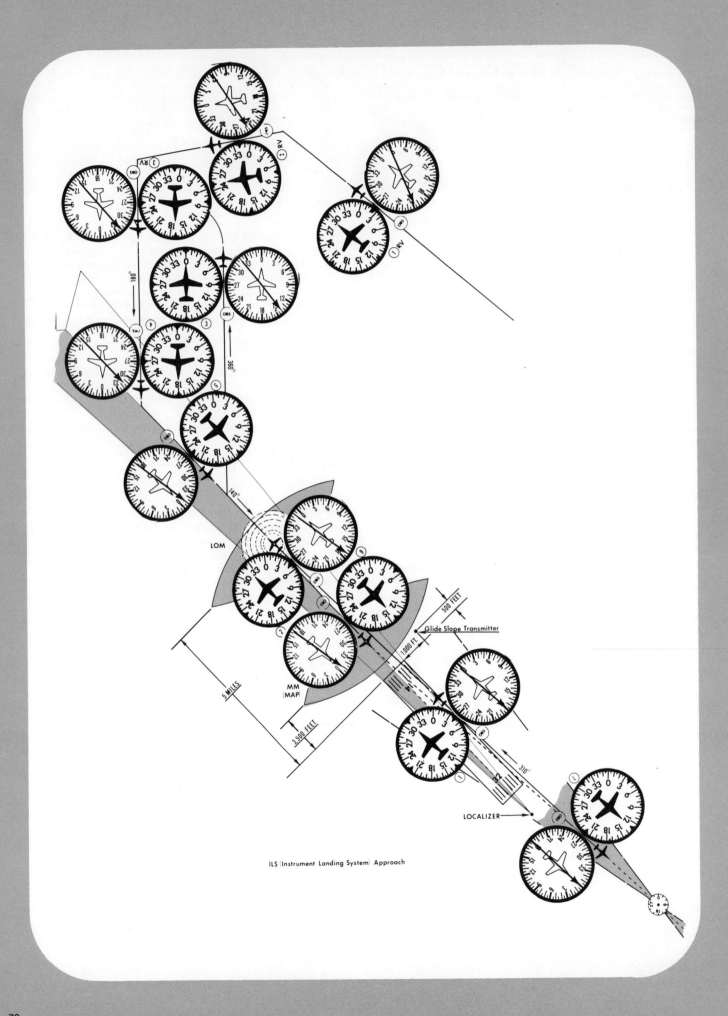

ILS (Instrument Landing System) Approach

PROCEDURE TURN

We will discuss the procedure turn first as we did with the NDB approach to show the similarity between the two. Then how that leads into the final approach (straight-in or circling) and the missed approach.

The procedure turn is the manuever prescribed when it is necessary to reverse direction to establish your aircraft 'inbound' on a final approach course. It is a required manuever except when the symbol NoPT (no procedure turn) is shown, when radar vectoring is provided; when a one minute holding pattern is published 'in-lieu-of' a procedure turn, or when a procedure turn is not authorized.

As with the procedure turn during an NDB approach the ILS procedure turn manuever must be completed in the prescribed area and on the headings noted on the approach chart.

The illustration to your left shows your flight path to the outer marker (LOM) is on the 310° bearing/radial FROM the NDB. Since the outer marker (LOM) does not radiate radials like the Omni, you must make sure you maintain an exact course, regardless of wind, FROM the NDB to the outer marker (LOM). Once the outer marker light lights up on your instrument panel and the ADF Direction Indicator needle swings to the 18 position — the tail of your aircraft — you begin your turn to the 'outbound' heading and maintain that for one (1) minute.

You've probably noticed by now the difference in the Course Indicator being presented here. This is the only type you can use for an ILS approach. You must have one needle (the vertical one) to show whether you're right or left of your course and the other needle (the horizontal one) to show whether you're too high (above the glide slope flight path) or too low (below the glide slope flight path). This is referred to as the 'fly down' or the 'fly up' position.

One reason that an ILS approach is more precise is the fact that the vertical and horizontal needles are more sensitive due to the frequency modulation that is built into the ILS system. That means that even the slightest deviation from either the horizontal or the vertical flight path will be noticed immediately. For instance, the horizontal dots on your Course Indicator when performing an ILS approach are representative of 1¼ degrees compared to 5 and 10 when using the ADF Direction Indicator needle for an NDB approach, and the vertical dots each represent ½ degrees. As an example; if the needle of your Course Deviation Indicator shows a one dot needle deflection when you are one mile from the touchdown point of your landing runway you are approximately 300 feet right or left (whichever side the needle is on) of the centerline of the runway. Enough to put you in the grass beside the runway unless you correct the error. Likewise a one dot horizontal needle deflection on your Course Deviation Indicator below the large dot in the center dial means that at one mile from your touchdown point you are approximately 25 feet lower than what you want to be, and that's not good.

The ground equipment required to give you this precision on an ILS approach consists of 1. the outer marker (LOM) and the middle marker (MM), which are called marker beacons, 2. a localizer transmitter located on the **opposite end of the approach end** of the ILS runway which helps you keep aligned with the centerline of the runway, and 3. the glide slope transmitter located 1,000 feet to the side and 500 feet from the end of the runway which helps you keep on the proper glide slope (angle of descent) to your landing runway.

At position #1 you're 'outbound' FROM the Omni on the 310° radial. At #2 you're 'on top' the outer marker (LOM) using your ADF Direction Indicator now and setting yourself up for the one-minute flight 'outbound' FROM the outer marker (LOM) and here the ADF Direction Indicator needle will swing to the 18 position — the tail of your aircraft — indicating passage over the outer marker. At #3 you're on the first turn of the procedure turn and after approximately one minute of flight (assuming no wind) you make a 180° left turn to intercept the ILS localizer path 'inbound.' Notice how far to the right the ADF Direction Indicator needle is a #4, yet you are ready to make your turn to intercept the localizer path 'inbound' when the ADF Direction Indicator needle swings to the 45° position (45° left to the nose of your aircraft). (THE COURSE DEVIATION INDICATOR NEEDLE IS MUCH MORE SENSITIVE WHEN USING THE ILS than it is when using the Omni. It is four (4) times more sensitive.)

At #5 you're 'inbound' and almost to the outer marker (LOM), your first clue as to how far away you are from the runway. At #6 you're at the middle marker (MM) and 3,500 feet away from the runway. This (#6) is also your missed approach point (MAP) and this is where you execute a missed approach if you cannot see the approach lights, runway lights or the runway.

Number 1RV, 2RV and 3RV are examples of what might transpire if you were radar vectored to the localizer course. Interception of the localizer course for a radar vector 'hands-off' is just like the (RBn) (NDB) radar vector EXCEPT THE ILS IS MUCH MORE SENSITIVE so you really have to be 'on your toes' when intercepting the ILS from a radar vector.

Position #7 simulates a missed approach.

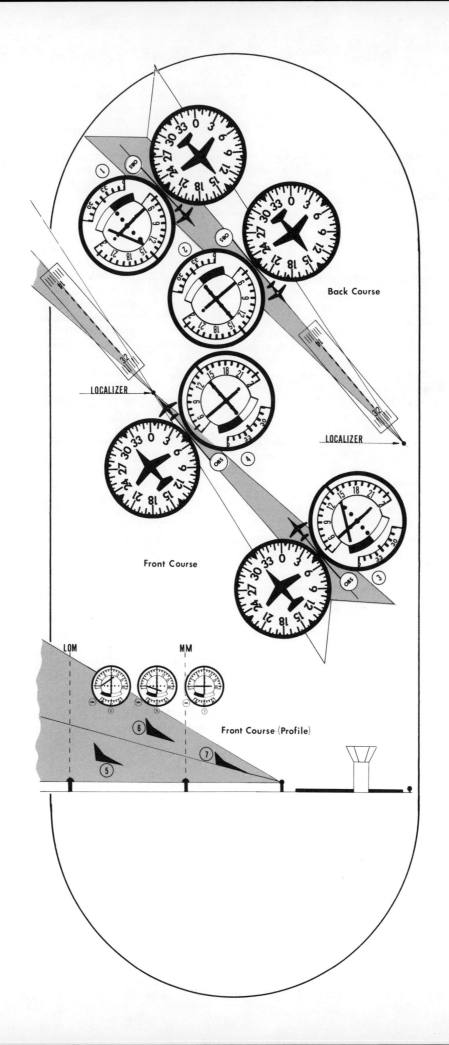

FRONT COURSE AND BACK COURSE APPROACHES

A FRONT COURSE APPROACH is an approach to an ILS runway that gives you vertical and horizontal guidance all the way to the missed approach point (MAP).

The reason the FRONT COURSE is a precision approach is that it employs so many more electronic aids than does the BACK COURSE (or the Omni or NDB approach) and allows very little deviation from the course, either up or down or sideways. The plan view of the FRONT COURSE approach shows the two marker beacons; the outer marker (LOM) which incorporates a non-directional beacon (NDB) too and the middle marker (MM), the glide slope transmitter to tell you whether you're too high or too low on your approach and the localizer transmitter which is, in reality, an extension of the centerline of the runway and lets you know when you're right or left of the centerline of the runway.

Contrast that to the BACK COURSE approach which has only the localizer transmitter as a reliable aid. The outer marker and middle markers are out of the picture because they are on the other end of the runway and only come into play when your aircraft passes directly over them. The glide slope transmitter is also on the other end of the runway. Another very important point to remember when using the BACK COURSE approach is that your Course Indicator is turned to the FRONT COURSE approach so therefore the old Course Deviation Indicator needle reversal syndrome pops up again. As long as you're aware of what's happening it's okay. BUT MAKE SURE YOU'RE AWARE.

Let's look for a moment at the profile view of the FRONT COURSE approach. Number 5 is too low, #6 is too high, and #7 is 'on glide slope' with the proper angle of descent.

The BACK COURSE approach is more like the Omni and the NDB approaches in that YOU MUST MONITOR YOUR RATE OF DESCENT ON THE ALTIMETER AND THE VERTICAL SPEED INDICATOR RATHER THAN THE GLIDE SLOPE NEEDLE.

The ILS, Omni and NDB approaches have MISSED APPROACH POINTS (MAP'S). The Omni and NDB missed approach points occur at the designated (published) Minimum Descent Altitude (MDA). The ILS missed approach point (MAP) occurs at the MIDDLE MARKER (MM). The same criteria applies to the ILS MISSED APPROACH POINT as it does to the Omni and NDB MISSED APPROACH POINTS; IF THE APPROACH LIGHTS, RUNWAY LIGHTS OR THE RUNWAY CANNOT BE SEEN A MISSED APPROACH MUST BE EXECUTED. And if, when performing a circling approach visual contact with the runway is lost A MISSED APPROACH MUST BE EXECUTED. And as before, DO NOT CONTACT AIR TRAFFIC CONTROL (ATC) UNTIL YOU HAVE YOURSELF AND YOUR AIRCRAFT UNDER CONTROL.

NOTE: The ILS missed approach point is at the Decision Height (DH) for a straight-in approach or at the Minimum Descent Altitude (MDA) if circling minimums are used.

THINGS TO REMEMBER WHEN FLYING ADF

1. When flying 'INBOUND' or TO an NDB know that a LEFT turn will always cause the ADF needle to move RIGHT and a turn to the RIGHT will always cause the ADF needle to move LEFT. THAT'S VERY BASIC, BUT VERY IMPORTANT.

2. When flying 'OUTBOUND' or FROM an NDB a LEFT turn will always cause the ADF needle to move LEFT and a turn to the RIGHT will always cause the ADF needle to move RIGHT. AGAIN — VERY BASIC, BUT VERY IMPORTANT.

3. After correcting for wind:
   a. If the ADF needle moves toward the nose of your aircraft (0) while tracking 'inbound', this indicates that your wind correction angle is NOT ENOUGH — take a greater wind correction angle and check the ADF needle after you are in wings-level flight to see if it was enough.

   b. If the ADF needle moves towards the tail of your aircraft (180), this indicates that your wind correction angle is TOO MUCH — in this case you can either reduce the wind correction angle or allow the wind to move you back on course.

4. When making ADF approaches
   After crossing the NDB, fly outbound for approximately two (2) minutes. Then start a left or right procedure turn at a 45° angle to the outbound course. (Note — when performing a LEFT PROCEDURE TURN, 225° and 45° are the 'magic' numbers and when performing a RIGHT PROCEDURE TURN, 135° and 315° are 'magic' numbers.)

IMPORTANT

When using the ADF you must constantly 'cross-check' the Heading Indicator with the Magnetic Compass. Heading Indicators have inherent precession error of one degree (1°) for every five (5) minutes of flight and three degrees (3°) for every fifteen (15) minutes of flight. That means that for every fifteen (15) minutes of flight you can be three degrees (3°) off course. And after one hour of flight, if you haven't corrected your Heading Indicator, you will be 15° off course. Too much for precision flying.

COMMON ERRORS IN THE USE OF NAVIGATION INSTRUMENTS
OMNI/VOR AND ADF

OMNI/VOR

1. Careless tuning and identification of station.
2. Failure to check the receiver for accuracy and sensitivity.
3. Turning in the wrong direction during an orientation (this error is common until you visualize position — not heading).
4. Failure to check the Course Indicator particularly during course reversals, with resulting 'reverse sensing' and corrections in the wrong direction.
5. Failure to parallel the desired radial on a track interception problem. Without this step orientation to the desired radial can be confusing. Since you think in right/left terms, aligning your aircraft position to radial/course is essential.
6. Incorrect rotation of the Course Selector (OBS) on a time/distance problems.
7. Overshooting and undershooting radials on interception problems. Factors affecting lead should be thoroughly understood, especially on close-in interception.
8. Overcontrolling corrections during tracking, especially close to the station.
9. Misinterpretation of station passages, on Omni/VOR receivers equipped without an ON/OFF flag, a voice communication on the Omni/VOR frequency will cause the same TO/FROM fluctuations on Course Indicator as shown on station passage. **Read the whole receiver before you make a decision.**
10. Chasing the CDI needle, resulting in homing instead of tracking. Careless heading control and failure to bracket the wind corrections makes this error common.

ADF

1. Improper tuning and station identification of station.
2. Dependence on homing rather than proper tracking, commonly results from reliance on the ADF indications instead of correlating them with heading indications.
3. Poor orientation, due to failure to follow proper steps in orientation and tracking.
4. Careless interception angles, very likely if you rush the initial orientation procedure.
5. Overshooting and undershooting predetermined magnetic bearings often due to forgetting the course interception angles used.
6. Failure to maintain selected headings. Any heading change is accompanied by an ADF needle change. **The instruments must be read in combination before any interception is made.**
7. Failure to understand the limitations of the radio compass and the factors that affect its use.
8. Overcontrolling track corrections close to the station (chasing the ADF needle) due to failure to understand or recognize station passage.